ⁱN 2012

Ŀ

2 1

KIM!

---

This ｉ

The ｉｃ
a furtʰ

*Also by Christopher Somerville*

Coast
Somerville's Travels
Britain and Ireland's Best Wild Places
Walks in the Country Near London
The Living Coast

CHRISTOPHER SOMERVILLE

# NEVER EAT  SHREDDED WHEAT

*THE GEOGRAPHY WE'VE LOST*
*AND HOW TO FIND IT AGAIN*

HODDER &
STOUGHTON

First published in Great Britain in 2010 by Hodder & Stoughton
An Hachette UK company

2

A CIP catalogue record for this title is available from the British Library.

ISBN 978 1 444 70463 1

Typeset in Baskerville by Palimpsest Book Production Limited,
Falkirk, Stirlingshire

Printed and bound in the UK by Clays Ltd, St Ives plc

Hodder & Stoughton policy is to use papers that are natural, renewable
and recyclable products and made from wood grown in sustainable forests.
The logging and manufacturing processes are expected to conform to the
environmental regulations of the country of origin.

Hodder & Stoughton Ltd
338 Euston Road
London NW1 3BH

www.hodder.co.uk

*In memory of Peter Mattinson,*
*my first and only Geography teacher*

# CONTENTS

# INTRODUCTION

## TRAVEL AND ARRIVE

There are no countries in the world less known by the
British than these selfsame British Islands.
– George Borrow, writer and walker, in his great gypsy book
*Lavengro* (1851)

Mr Mattinson was one of my teachers when I was ten
years old. He taught me Maths and Geography, my two
worst subjects (I hadn't encountered Science yet). If I hated
Maths, I loathed Geography with a passion. I just couldn't
understand it, probably because there were better things to
think about on the day we had our first Geography lesson –
perhaps one of those plastic DIY assembly toys out of the
cornflakes box that I used to fiddle with under cover of my
desk lid. Whatever the reason, I obviously missed out on
some vital blast of facts, right at the start. North, east, south,
west: that kind of thing. From then on, Geography was, quite
literally, a closed book to me.

Mr Matt tried his best, but the child-friendly analogies he
employed made no sense to this ten-year-old. One image in
particular baffled me for years. 'The British Isles,' said Mr
Matt, 'look like an old hag with a big nose, wearing a bonnet,
shouting at a parrot and riding on a pig.' *What?* 'Oh, yeah!'
chortled the other kids, gleefully pointing out the hag and
the pig to each other. *Eh?* I stared at the outline on the wall
map, but I just couldn't see hag, pig or parrot. Ten years

later, sitting cross-legged on the floor of a student flat and gazing at the cover of a Philip's School Atlas on which a friend was attempting a little home-grown horticulture, I suddenly got it. Oh, yeah, man, I see . . . Hag on a pig, yeah! Parrot, yeah! Wow, far out!

The hag, the pig and the parrot weren't the only things to puzzle me about Geography. There was also the tricky business of Shredded Wheat. When faced with map or atlas, kids would murmur, in a pensive but knowing way, the mantra: *Never. Eat. Shredded. Wheat.* I didn't get that until I embarked on a short-lived career as a teacher, twenty years later. 'North, East, South, West, sir – the initials, sir,' explained nine-year-old Neil from the back row. Ah. 'Round the compass like a clock, sir.' Mmm.

By the age of thirteen I'd abandoned Geography as a boring and baffling stew of facts. Boring and baffling not just to me, but to most of my contemporaries – and to my children, and *their* contemporaries, too, if I can believe what I'm told. Too many moraines and rift valleys, too much banging on about coal tar by-products and the fluorspar industry. Enough already with imports and exports, changes in manufacturing employment in the Vale of Belvoir, square hectares of Renfrewshire under root crops. Too much confusion of conservation, ecology, world population movements, climate change and other doomsday scenarios. And not enough of the basics, the stuff everyone really wants to know: what's where, how to get there, and what it's like when you do get there.

Those are the practical advantages of knowing one's Geography. But there's another benefit, just as important. Geography helps me notice and appreciate what's around me, while I'm on the way to where I'm going. It puts a polish and a sparkle on things. 'To travel hopefully', said clever old

Robert Louis Stevenson, a world-bestriding traveller himself, 'is a better thing than to arrive.' What a fat old cliché! And how fabulously true! The British are lucky enough to live in just about the most diverse, compact, beautiful and endlessly fascinating set of islands in the world. To have the privilege of travelling them – which we endlessly do, for work or play – is to move through the world's best art gallery, nature reserve, library, concert hall and theatre – all for free.

*

So why don't we see what's there any more? What's blocking us off from our birthright of Geography? Is it bad teaching of the basics? *No* teaching of the basics? Because we are all somehow thicker than Robert Louis Stevenson and his generation? Half a dozen theories . . .

1. Children don't get out and wander about their local streets and countryside as they used to, because this generation of parents, bewitched by health and safety, harbours irrational fears of traffic, or stranger danger, or accidents by flood and field. Therefore children never learn to absorb the landmarks, unimportant in themselves – a tree, a gate, a bend in the lane – that make up their own personal Geography.
2. We are all frantic to get where we're going as quickly as possible. Work pressures, social arrangements, I-can-be-there-for-the-meeting-if-I-leave-home-at-4 a.m. – our fast cars and 125 mph trains and Edinburgh-in-forty-minutes planes force the pace, and we blindly follow.
3. We don't need to look out of the window at the outside world, because the outside world is now inside the car or the bus or the train carriage with us: the boss on the

mobile, the stock market on the dashboard Internet, the email bleeping on the BlackBerry, the news on the laptop.

4. In spite of being more contactable by the outside world, we are also more insulated from it. What do the rainy hills, the budding trees or the sun-dried fields, the smell of the earth or the crunch of a frozen puddle have to do with the cosy, cut-off world we inhabit as we flash by – a world whose sounds, smells, climate, light and shade we can select to suit ourselves at the press of a button?

5. And that applies to foot and bike travellers, too, iPods plugged in, shades on, insulated by Gore-tex and Neoprene, pumped up by adrenaline advertising, staring ahead and burning calories, using the countryside as a gym – To The Max! Go For It! Rippin' Up The Ridgeway!

6. Planning a journey, and then doing it, have been reduced by GPS, Sat Nav, Google Maps and other positional and directional tools to a matter of (a) where to start and (b) where to finish. Everything in between is taken care of by 'someone else' – namely, the little personal servant goblin who lives in the gizmo and tells us exactly where to turn left and how far it is to the next service station. So we read Ordnance Survey maps and road atlases less; we have less peripheral context at any given place, because we're missing the wealth of superfluous, but civilising and enriching, detail inherent in maps, so plump with facts and knowledge, so redolent of our huge heritage of national culture and history.

To move through a GPS landscape of grey blanks knitted together by blue, green, orange and white spider lines is to negate the very notion of Stevenson-style travelling. Lay the Google and OS 1:25,000 Explorer maps of the Stonehenge area side by side. On Google – roads and a ghostly hint of buildings. That's it. On the Explorer, all round the mighty henge itself: ridged and billowing downland, ancient trackways, processional paths, long barrows and tumuli where our distant lordly ancestors lie buried, the mysterious banking of the Cursus track, copses and spinneys bounded by unexplained earthworks.

Do we actually *need* all this stuff to get from Amesbury to Winterbourne Stoke? No, we don't. Should we delight in it, and feel grateful to be part of it, and smack our imaginative lips over it, and be inspired to come back and explore it with a flower book and an archaeology book on a sunny day soon? Absolutely. That's our national birthright.

*

Here is a book for anyone who:

- Doesn't have a clue about the basic Geography of the British Isles
- Once knew all, or some, or a bit of it, and has forgotten it, or lost it, or let it slip
- Goes away for football matches, or music festivals, or holidays
- Goes walking, or riding, or biking at a slow pace through these islands
- Catches themselves on the motorway or the railway, dashing past a tower in a wood, or a range of blue hills, or a lake behind rushes, and thinking: 'Where is that? And why's it there? And how can I find it again?'

Here you'll find, placed and explained, the West Country and the North Country and the South Country. The islands and Highlands, the roads and rivers. Lumpy bits and bumpy bits, fens and flats. Cities and towns you know, or would like to know, or haven't given a damn about up till now. The moors and mountains, coal pits and quarries. Landmarks. Waymarks. The best-known and the least-known parts of these patchwork, magical islands of ours, rediscovered and plainly set out, so that anyone – even a geographical donkey such as myself – can enjoy the journey and everything that's along the wayside.

# THE BASICS

*North, East, South, West*

There are four main points that we take our directions from: **N**orth, **E**ast, **S**outh, **W**est.

To remember where the four points are, just picture a compass and a clock.

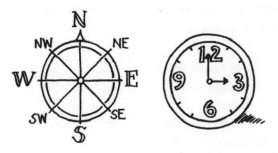

The compass is laid out exactly like a clock.

At the top (12 o'clock) is **N**orth. On the right, at 3 o'clock, is **E**ast. At the bottom, at 6 o'clock, is **S**outh. On the left, at 9 o'clock, is **W**est. N, E, S, W ... **N**ever **E**at **S**hredded **W**heat.

Take any place on the map – for example, London. Whatever's above it is north of it (e.g. Stevenage); whatever's to the right is east of it (e.g. Southend-on-Sea); everything below it is south of it (e.g. Brighton), and everything to the left is west of it (e.g. Reading).

*What about the bits in between?*

Halfway between north and east is north-east (7½ minutes past the hour).

Halfway between east and south is south-east (22½ minutes past the hour).

Halfway between south and west is south-west (22½ minutes to the hour).

And halfway between west and north is . . . north-west! (7½ minutes to the hour).

Example: Travelling out of London, heading north-east, you would reach Chelmsford, south-east Tenterden, south-west Guildford and north-west St Albans.

# THE BRITISH ISLES

This island is almost made of coal and surrounded by fish.
— Aneurin Bevan, Welsh politician

A traveller in the upper stratosphere, passing over the British Isles, sees a small, fantastically ragged archipelago (cluster of islands) sprawling in the sea just off the north-west coast of Europe. Two main chunks of land catch this space voyager's eye. There's mainland Britain, its west coast bitten by the incessant Atlantic waves into deep slices and jagged peninsulas, its eastern shores rounded and smoothed by the hungry tides of the North Sea. And there's the island of Ireland lying to the west, with the action of the seas producing just the same effect – deeply fissured western seaboard, straighter and smoother east coast. All around the two main islands, but mostly off their west and north coasts, hundreds of smaller islands lie scattered.

All the islands of the archipelago, big and small, look green. They are lumpy with hills and mountains, blue-veined with rivers, edged with shores of grey, black, red, white and yellow rocks, sands and shallows. Exposed around these coasts, forming the mountains and underlying the green lowlands, is a riot of rock, hard, soft, rigid, squashy – every kind, from some of the oldest rocks in the world to some of the youngest. Hard-edged shadows of cities are concentrated here and there, especially across the waist of mainland Britain. If the traveller high overhead were to take his flight by night, he would stare down on an inverted star map, the cities like brightly lit ships riding amid dark seas. By day or night the beauty is breathtaking, the variety mind-boggling. The archipelago lies isolated, but also protected, by its surrounding sea.

> This precious stone set in the silver sea,
> Which serves it in the office of a wall
> Or as a moat defensive to a house,
> Against the envy of less happier lands, –
> This blessed plot, this earth, this realm, this England.

William Shakespeare cited 'England' in that famous passage in *King Richard II*, but it is the whole sea-surrounded archipelago of the British Isles that his fabulous image evokes.

## Mainland Britain

Think of a triangle, and you've got the rough picture. Scotland (the hag's nose, hairdo and bonnet) is the northern bit of the main island, at the top, with its south-western corner looking across the sea at Northern Ireland (the parrot's head). West-facing Wales (the pig's head) lies on the left of the main island, looking across the sea at the Republic of Ireland. All the rest is England.

### England

**England** fills all of the south and most of the east of the British mainland. Shakespeare's 'blessed plot', the largest of the three countries that make up mainland Britain, is also the most prosperous. That's largely thanks to its underlying rocks, the remarkably varied geology that shapes its landscape and has shaped its history, too. England has less wild, rough land than mountainous Wales and Scotland; it is

softer, gentler, more easily farmed. It's a small country, too – you could very comfortably drive from bottom to top of England in a day without breaking the speed limit. But it has a hugely long coastline for its size – it would take you a fortnight to drive around the edge of England at the same speed. That coast is deeply indented by the sea, and it has hundreds of harbours and ports, both great and small. England has also been incredibly lucky in having huge deposits of precious minerals like coal, iron, tin, copper and quarry stone that sparked its Industrial Revolution; and in possessing the tumbling streams that powered the mills and factories, and the wide rivers that transported what they made long before the rest of the world caught up.

Five out of six people in the British Isles live in England, with roughly one-third of these living in London and the south-east (bottom right). There's a good reason for that: the south-east part of England, with its geology of chalk and clay, is the driest and warmest part of the entire island, and it's the closest region to Continental Europe, too. So it's no coincidence that it contains Britain's capital city of **London**. Dry, warm and easy to reach sounded good to the invading Romans 2,000 years ago, when they planted up the south-east with corn and grapes – and the south-east of England is still the most prosperous, and pricey, part of Britain. North of (just above) the south-east corner of England, clay-bound **East Anglia** (easily recognisable by its rounded, east-facing coastline) shares the same climate – warm in the summer, cold in the winter, and generally dry. Here is the best farm-land in Britain, a lot of it rich black peat lying below sea level, protected from the North Sea tides by sea walls and complicated systems of drains and sluices. The great apron of land at the eastern edge, known as the Fens and absolutely dead flat, was once the bed of the sea, from which men reclaimed it. Across in the opposite corner of southern England, the south-west (bottom left), sometimes known as the **West Country**, has a mild climate, too, but is a lot wetter and greener – good dairying country based on fertile lime-stone and sandstone. This is the classic English seaside holiday region, with the gentle tides of the English Channel for bathing and the crashing rollers of the Atlantic for surfing. The West Country coast of red sandstone or darker, harder granite is pounded by the waves into coves and bays floored with soft sand.

The further north and west (upwards and left) you go in England, the wetter, greener and hillier the countryside becomes; colder and harsher, too. There's a blotch of industry and a tangle of canals near Shakespeare Country, in the centre of the **Midlands** around Birmingham, and another one to the north between the western seaport of Liverpool and the triangle of Sheffield, Manchester and Leeds, former manufacturing giants that owed their prosperity and huge, sudden nineteenth-century growth to the coal and iron that underlie them and to the dampness of their valleys, ideal for cotton-spinning. North again it's the wide landscapes of **Yorkshire**, England's largest county, with its green dales (valleys towards the west of the county, cut out of soft lime-stone and harder gritstone by rain and rivers) and dark moors (north-east Yorkshire's domes of acid sandstone topped with peat and heather).

Country walkers adore wide-open Yorkshire, but the real walker's paradise is up in the far north-west of England (top left corner), where a jumble of ancient slate, volcanic rock, granites and sandstones has built the high and beautiful fells (tall hills) of the **Lake District**, beloved of poets and hill walkers. Here you'll find England's highest mountain, the 3,209 ft/978 m Scafell Pike. The industrial landscape of the Rivers Tyne and Wear, which lies in the **North-East** (top right), used to be famous for coal-mining and ship-building until those industries disappeared late in the twentieth century – now it, too, is reinventing itself as walking and cycling country. Right up in the **Scottish Borders**, between England and Scotland, it's rolling, hilly land with lots of dark conifer forests, and a hauntingly lovely coast of cold seas and long, windy beaches.

There's a long-running rivalry, part joke and part reality, between the south and the north of England. Southerners (folk who come from anywhere south of the Midlands) caricature those in the north as uncouth, blunt, uncivilised, and ready to steal your hubcaps. To northerners, there's nothing more feeble, pathetic and namby-pamby than a 'soft southerner'. There might be something in both viewpoints.

## Scotland

North of (above) England sits **Scotland**, iconic land of red deer, heather, mountains, bagpipes, kilts and whisky – to evoke but half a dozen 'corny-but-true' images. A bigger mass of land than most non-Scots suppose, Scotland and its hundreds of islands account for almost one-third of the main island in bulk, and well over half its coastline. If there's one image of Scotland that holds sway in everyone's imagination, it's of majestic mountains interspersed with deep, sweeping glens; if there's another, it's the magically misty and beautiful islands of the west. The romanticising of the tragedy of Bonnie Prince Charlie had a lot to do with that. So did the notorious Highland Clearances of the nineteenth century, when the clansmen were swept from their native glens and

replaced by more profitable sheep, leaving the magnificent landscape empty and silent.

The east of Scotland is drier than the Atlantic-facing west, which catches the lion's share of the rain. The east is also far less mountainous, and less indented by the sea, so communications are much easier – and that's why, as with England's capital London, Scotland's capital city of **Edinburgh** is sited in the south-east of the country, and has become the hub of a self-confident resurgence of Scottish nationalism in recent years.

Scotland is traditionally divided into **The Lowlands**, the more southern part, and **The Highlands** further north – self-explanatory names, with the lower ground tending to be in the Lowlands, and the higher ground . . . well, you work it out!

The Lowlands extend north from their border with England as rolling, billowing country of hidden valleys, high wild hills and fertile river valleys, where neat little towns lie widely separated. There's a belt of coal and iron around the waist of Scotland between **Glasgow** and Edinburgh, with a sprawl of industrial towns. North again is increasingly hilly country; generally speaking, the further north and west (up and left) you go in Scotland, the wetter and hillier things get. Just as in England, in fact. The Lowlands extend north as far as the Great Glen – the ruler-straight valley containing Loch Ness that runs from Fort William in the south-west (near the bottom left corner of Scotland) to Inverness in the north-east (near the top right).

The Highlands include everything north and west (above and left) of the Great Glen. They also make a bulge into the north-west part of the Lowlands to include the highest bits of the great Grampian Mountain range. The west coast is slashed into tatters by the ceaselessly pounding Atlantic, and overlooked by a superb succession of mountain ranges. Up in the far north, things level out into the vast, million-acre boglands of the Flow Country, before reaching the chilly, stormy seas and ancient, iron-hard quartzites, granites and sandstone of Scotland's (and mainland Britain's) most northerly coast.

## Wales

Land of song, poetry and romantic mythology, **Wales** lies to the west (left) of England. Wales is roughly square in shape, with three of its sides – north, west and south – bordered by sea, and its western coast deeply scooped into a concave arc. The east side connects Wales to England with a wriggling, meandering border that wanders roughly south–north through glorious hill country.

As with both England and Scotland, the east and south of Wales face well away from the rain-laden winds off the Atlantic Ocean, and tend to be drier, warmer and less hilly than the north and west. Here is where the Welsh capital city of **Cardiff** is situated, at the feet of the deep, southward-curving Valleys of South Wales. An astonishing wealth of coal, iron, copper and other minerals hereabouts made the Valleys the powerhouse of Wales during the Industrial Revolution, but they lie quiet in these post-industrial days.

Wanderers looking for a superb long-distance coastal walk, a sandy seaside holiday or a trip to bird-haunted islands make for Pembrokeshire, out at the tip of south-west Wales – a corner of the country that's warm, wet and sandy of coast. West Wales, based around the long, west-facing coast of Cardigan Bay, is the stronghold of Welsh language, poetry and culture. The intensely rural and hilly region of **Powys**

fills the centre of Wales, a land of small market towns, conifer forests and high hill ranges reflected in dozens of lakes and reservoirs. Ancient drovers' roads criss-cross the eastern (right-hand) border of Powys, where the eighth-century earthwork of Offa's Dyke carries walkers on a roller-coaster journey along Wales's national boundary with England.

The north-west of Wales, as with both England and Scotland, is the wettest and most mountainous district, and here at the heart of Snowdonia stands Wales's highest mountain, kingly Snowdon itself, surrounded by a court of jagged, formidable peaks of slate, granite and baked volcanic rock – a landscape quarried, dug and mined, but somehow transcendent.

## Northern Ireland

From its perch in Ireland's north-east (top right) corner, **Northern Ireland** looks east towards mainland Britain across the Irish Sea. If any part of the British Isles deserves the hackneyed old labels of 'forgotten' or 'undiscovered' or 'best-kept secret', it's this. The Republic of Ireland is separated by a wandering border from Northern Ireland, a region roughly circular in shape, with enormous Lough Neagh – the largest lake in the British Isles – at its heart.

The coasts of Northern Ireland are some of the most dramatic and beautiful in these islands, especially the north- and east-facing Antrim Coast, where enormous volcanic cliffs lead to the famous basalt promontory of the Giant's Causeway. The Atlantic and Irish Sea have carved deep into the volcanic coasts of Northern Ireland, producing a series of picturesque sea loughs or inlets, from Lough Foyle in the north-west round to Carlingford Lough in the south-east. The capital city of **Belfast** lies at the head of one of these,

Belfast Lough on the east coast, with its easy access to mainland Britain.

Just as it says in the old and oft-quoted Percy French song, the 'Mountains of Mourne sweep down to the sea' in the south-east corner of the region. Beyond, in the south around the border, lie scores of rounded little hills, tight-packed and cosy-looking, the grassed-over gravel hummocks or drumlins left behind when the Ice Age glaciers retreated 10,000 years ago. The fine walking uplands of the Sperrin Hills rise west of Lough Neagh, while down in Northern Ireland's own 'Lake District' of **County Fermanagh** in the south-west curves the great lake system of Upper and Lower Lough Erne. Were our putative stratospheric traveller to glance down as he crossed this watery Fermanagh landscape, he would – if he had any poetry in his soul – see a great shining dolphin leaping north-west and scattering a burst of spray in its wake.

# 2

## LAND AND BORDERS – ENGLAND

Where are the Fens? 'Well, you know you're here when you're here,' was a Fenman's description of his native territory when I put the question in a Wisbech pub. 'And you know you're out when you're out.' Hmmm, yes, I sort of saw what he meant. But it wasn't all that helpful.

There are a bewildering number of names for the many regions that make up England. And a load of questions, too. Which are the borders of the West Country, or East Anglia, or the Lake District? Where are the Fens, or the Yorkshire Wolds, or the Welsh Marches?

You could argue for a long time about names and precise locations of the different parts of England, Wales, Scotland and Northern Ireland. The good old-fashioned counties have been reorganised and re-reorganised until everyone's in a nice stew of confusion. You live in Bristol, right? But are you sure it isn't actually somewhere called Avon, that used to be Somerset? Or is that Somerset again now? No, it's South Gloucestershire. Or is Bristol a county all by itself . . . ? Unless it isn't . . . ?

Let's cut through all that.

Starting down in the south-west, here are the ten regions of England, and the counties they now contain:

## West Country
(Cornwall, Devon, Somerset)

The West Country is for holidays. Why else would the long south-westerly toe-tip of England be blessed with so many wonderful sandy beaches, sheltered smugglers' coves, pounding surfers' waves, cream teas and pints of cider? Not to mention the world's greatest rock music festival, Glastonbury, and a laid-back air that other, more uptight regions can only envy. Add in the geology of the peninsula – limestone to produce those fertile dairying pastures, sandstone for the red cliffs and the heathery uplands of **Exmoor** National Park, granite that outcrops in the dramatic tors (wind-sculpted rock piles) of **Bodmin Moor** and **Dartmoor** National Park. Top it all off with twisty, high-banked lanes, ancient thatched cottages built of cob (straw and clay), and a broad local accent that burrs and rumbles seductively. It must be chilled-out heaven to live there.

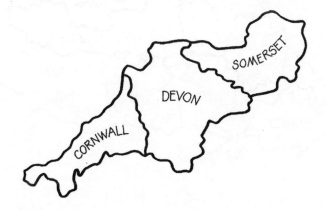

Well ... there's more to it than that. That Cornish granite is tough stuff. The moors of **Cornwall** hold the ruins of hundreds of tin mines, harsh places of dangerous work and little pay; farming is sparse and poor, on small fields exposed to salty winds; fishing is hazardous work. The lush-looking dairy farms of **Devon** and **Somerset** struggle with low milk prices. Most country communities have to cope without buses, schools, pubs or shops. That West Country blue sky has a few clouds across it, for sure.

But Cornwall, with the warm tides of the Gulf Stream moving offshore, has the UK's best weather, the earliest flowers and the best sea bathing. The beauty of Devon and Somerset, their red earth and green fields, the wild red deer roaming Exmoor, the cider apple orchards – they're all still there, down in the south-west, if you get off the main roads and go looking for them.

## South Country

(Wiltshire, Dorset, Hampshire, Isle of Wight, West Sussex, East Sussex, Kent)

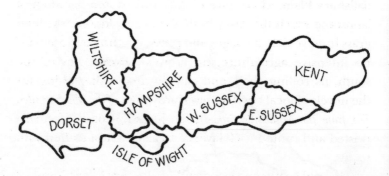

The South Country contains the southern English counties that lie west to east along a single, mighty rampart of chalk. This extraordinary stuff, five times the height of Big Ben at its tallest, is made entirely of the shells and body parts of tiny sea creatures which lived and died in the Great Chalk Sea, a shallow tropical sea that covered most of Britain and neighbouring parts of Europe some 60 million years ago – around the time the dinosaurs died out. Chalk eroded by the sea forms the famous milk-white cliffs of the south coast; chalk moulded by wind and weather underpins the billowing shapes of the **South Downs**. It's easily cut by a plough, full of lime for the crops, packed with flints for tool-making. No wonder men settled here as soon as they could when the Ice Age glaciers retreated; no wonder the South Country counties are so rich in the hill forts, stone tombs and earthworks of our distant ancestors.

Moving from west to east along the chalk belt, you start with the landlocked county of **Wiltshire**, dotted with antiquities (mighty Stonehenge, Avebury village trapped in its double ring of ancient standing stones, White Horses cut in the chalk hillsides) and centred on the vast emptiness of **Salisbury Plain**, where rare wildlife thrives in a paradoxical landscape which the Army both abuses (explosions, shellfire, grass fires, vehicle damage) and preserves (no development, no intensive agriculture, no crowds). **Dorset** lies to the south, its rolling chalk and clay downland the setting for the novels of local boy Thomas Hardy (*Tess of the D'Urbervilles, Far from the Madding Crowd*), its 'Jurassic Coast' of crazily twisted and canted rock layers a treasure chest stuffed with fossils.

The wild heaths and woodlands of the **New Forest** National Park join Dorset to its eastern neighbour of **Hampshire**, the halfway point of southern England's coast. Hampshire's coast is flat and soft, a marshy, creek-cut tatter of waterways and small peninsulas, where the uneasy sister ports of Portsmouth and Southampton (bitter football rivalry ahoy!) lie sheltered by the diamond-shaped **Isle of Wight**. Inland, the roller-coaster hills of the **South Downs** National Park flow east from Hampshire into **West Sussex** and on across **East Sussex** into **Kent**; classic downland, smooth and gentle, whose heights are threaded by walkers on the South Downs Way National Trail. The coast, by contrast, is a series of splendid white cliffs of chalk, like teeth bared at the sea, studded with seaside resorts such as Brighton, Eastbourne and Margate, all the way east from East Sussex round to the muddy shores of the **Thames Estuary**.

## Home Counties

(Surrey, Berkshire, Buckinghamshire, Hertfordshire)

Chelsea tractors in every lane? Check. Hugely expensive houses? Check. Private schooling, private stables, gymkhana rosettes, guard dogs, helicopter pads? Check. Foot-ballers' wives tipping back the Chardonnay behind electric gates; rock stars retired disgracefully to their country piles? Yeah, check, baby. But that's only how *some* of 'em live, out in the Home Counties. Railways, motorways, dual carriage-ways connect the 'gin-and-tonic belt' umbilically to the capital. **Surrey**, **Berkshire**, **Buckinghamshire**, **Hertfordshire**

– they encircle London closely on the south, west and north; they get their sheen and their polish of wealth and good living from a commuting way of life. Real farming, real local industry, real life do go on in the four Home Counties (or five – some would include Kent), but sometimes it's hard to believe.

Running clockwise round the M25: Surrey has the **North Downs** and the beech woods; Berkshire has the sandy heaths, the pine trees, the aristocratic and royal connections (Windsor Castle, Eton College, Sandhurst Royal Military Academy). 'Leafy Bucks' and its eastern neighbour Hertfordshire share the beautiful **Chiltern Hills**, a rounded rampart of chalk that curves north-east (from 9 to 12 o'clock) beyond the far outskirts of London. Here are the classic bluebell groves, the beech woods and pretty chocolate-box villages that Londoners picture when they dream of living 'out in the sticks'.

## East Anglia
(Essex, Suffolk, Norfolk, Cambridgeshire, Lincolnshire)

'Very flat, Norfolk,' yawned Noël Coward in *Private Lives*. That's right, isn't it? Actually, no. Noël was sacrificing the truth for a good line. Quite rolling, Norfolk, as is most of the rest of East Anglia. But there *are* flat bits in this easternmost region of Britain – spectacularly flat prairies that shoot off to the horizon with never a bump or pimple. East Anglia feels rather apart from the rest of England; with its rounded coast profile, it sits above London, but is bypassed by all the motorways except the M11 (and even that peters out at Cambridge, as if it can't be bothered to go any further). Writers, painters, musicians, saltwater sailors and other city escapees tend to choose East Anglia over the Home Counties, because – well, to be frank, because it's less precious, more down-to-earth, and cheaper.

Much-maligned **Essex** ('nobody loves us and we don't care!') runs east from London in a sprawl of ribbon development towns. Ugly, boorish, scruffy, flashy, trashy – Essex has heard it all. Bits are like that, especially around the ribbon towns. But most isn't. Forgotten forts and wildlife marshes lie among the oil jetties and landfill sites along the Essex bank of the **Thames Estuary**, a shore with the moody charm of a place no one visits. Stretching north from the river mouth for fifty miles is a splattered coast of flat islands, muddy creeks and marshes, all but deserted. Inland Essex begins to roll the further north you go, a land of dipping cornfields and ancient woods, quite content to be nobody's place of resort.

The 'heavenly twins' of East Anglia are **Suffolk** and **Norfolk**, filling the region's round easterly bulge, Suffolk ('Land of the South Folk') under Norfolk ('Land of the North Folk'). Sheep-rearing, wool-trading and cloth-working in the Middle Ages brought the wealth that built the region's famous flint churches, the finest in Britain. Internally the land is heavy clay, great for corn-growing, with a rolling motion that flattens out as you journey north over the Suffolk/Norfolk border into more chalky claylands. A strip of country just east of Norfolk's capital city of Norwich is full of flooded peat diggings – the celebrated cruising lakes of the **Norfolk Broads** National Park. Out by the sea it's mostly flat, with marshy creeks and pebble beaches, except for a section of the North

Norfolk coast bounded by very crumbly cliffs of sand and clay. North Norfolk's other claim to fame is its salt marshes, which have grown more than a mile out from shore and choked off the sea trade of a string of handsome little villages. Wildfowlers and birdwatchers love this area.

**Cambridgeshire**, on the western borders of Norfolk and Suffolk, fulfils Noël Coward's prejudice – it *is* flat. Here, stretching in a vast arc towards the sea through Cambridgeshire and Norfolk, is the area called **The Fens** – though very few patches of true fen (reed swamp and wet woodland) remain. Almost all was drained for agriculture, as were the shores around the big Wash estuary. Roses and narcissi, carrots and spuds grow wonderfully in the peat and silt. Further north runs Lincolnshire, with its dead-flat sea margin and its humpy spine of hills, the **Lincolnshire Wolds**. Like Essex, Lincolnshire is under-visited and under-appreciated, a county for connoisseurs who like to take their time.

## South Midlands

(Gloucestershire, Oxfordshire, Warwickshire, Worcestershire, Northamptonshire, Bedfordshire, West Midlands)

The Midlands lie like a great lump in the stomach of Britain, with an industrial belt around their middle. South of that Birmingham–Coventry–Leicester line, the South Midlands are still recognisably part of southern England, with their abundant oak woods, half-timbered houses and gently rolling grasslands. Ironstone and limestone quarries speak of past industrial activity, but mostly this is quiet 'middle England' to the hilt.

The two side-by-side counties of **Gloucestershire** and **Oxfordshire**, filling the southern part of the South Midlands, are knitted together by a seam of beautiful golden oolitic limestone that forms the 'horse-and-hounds' country of the **Cotswold Hills**. North of here you get to **Warwickshire** and the market town of Stratford-upon-Avon – and you know who was born there, don't you? A gazillion American and Japanese tourists do, that's for sure. 'William Shakespeare Country' lies squeezed between **Worcestershire** on the west (**Malvern Hills**, Edward Elgar) and **Northamptonshire** on the east (the M1, the Grand Union Canal, and, um, er ...). Then **Bedfordshire**, east of that (lots of flat clay, Milton Keynes, and, er, um ...). Some portions of England, such as the two last-named counties, seem to dive beneath a casual visitor's radar. You'll just have to go there and seek out Salcey Forest and Twywell Gullet, Bunyan's Oak and the Icknield Way for yourself. It's worth it.

North of Warwickshire, in a built-up smear where the M5 and M6 meet, lies the big conurbation of the **West Midlands** – in essence, the city of **Birmingham** and the spider's web of long-established industrial villages on its north-west flank that goes by the name of the **Black Country.** Coventry to the east shares the same kind of character – noisy, grubby, multiracial, lively, great music, very funny people.

## North Midlands

(Staffordshire, Derbyshire, Nottinghamshire, Leicestershire, Rutland)

Sandwiched between the industrial sprawls of the West Midlands and the Liverpool–Manchester–Sheffield belt, you'll find the cluster of counties that make up the North Midlands. This is where hard northerners sense they are sinking into the decadent South, while southern softies, feeling themselves 'up  North', see with their own eyes the dereliction caused by the collapse and disappearance of heavy industry. Above Birmingham sits **Staffordshire**, where clay, coal, iron and limestone (perfect for the now redundant pottery-making industry) underlie the **Potteries** area. Here vast swathes of derelict clay pits, kilns and mines are being massively redeveloped for housing.

East of Staffordshire lies **Derbyshire**, a county dominated by the focal point and symbol of the North Midlands, the **Peak District** National Park – beautiful walking country, with limestone dales in the south (the 'White Peak') and gritstone moors in the north (the 'Dark Peak'). East again it's **Nottinghamshire**, part industrial (coal mines, textiles, engineering – most of it now gone), and part rural (**Sherwood Forest**). South of Nottinghamshire run the broad acres of **Leicestershire**, home of the ancient and world-famous Quorn Hunt – a classic fox-hunting, horse-galloping county with its big, flattish fields. And don't forget quiet little **Rutland** out on the eastern fringe of the region, famous for nothing except being England's smallest county – it's the same size as the Isle of Wight.

## Welsh Borders

(Herefordshire, Shropshire, Cheshire)

Lying along the eastern edge of Wales are three English counties known as the Welsh Borders – sometimes called the **Welsh Marches**. Down in the south it's **Herefordshire** of the red earth, green cattle pastures and cider apple orchards. In the middle it's **Shropshire**, a county that rolls and lifts the further westward one goes towards the rounded uplands of the **Long Mynd** and the deep Border valleys. The northernmost Welsh Borders county is **Cheshire**, another of those counties people don't really have a strong image of – flat in the west, industrial where it touches Merseyside, lake-bespattered in the north, hilly in the east as it rises to meet the Peak District. And full of Manchester United footballers' country piles.

## Yorkshire

(East Riding of Yorkshire, South Yorkshire, West Yorkshire, North Yorkshire)

A whole region composed of just one county? Yes, I know . . . but this is Yorkshire. 'Best bloody county in the country,' is the opinion you'll get if you ask a Yorkshireman. It's certainly the biggest, and the most varied, and accounts for a good half of what's often called the North Country. Not content with having two National Parks all to itself, Yorkshire pinches the top slice of the Peak District, too. You could fit forty Rutlands into Yorkshire. So it's quite some county.

There are four parts to Yorkshire:

**The East Riding**, a flattish sweep of country apart from the gentle rise of the **Yorkshire Wolds**, this is (as you might guess) the eastern part of the county. The broad River Humber washes past the port of Hull on the East Riding's southern flank, and the North Sea eats away at its eastern section (also known as Holderness) – crumbly black cliffs and sensational, deserted beaches.

**South Yorkshire** is the smallest and traditionally the most heavily industrial part, where the M1 and M18 meet. Sheffield and Rotherham make iron and steel, Barnsley and Doncaster produce coal and brass band music – everyone knows that. Well, everyone's wrong, everyone who hasn't been there in the past few years. The coal industry's just about dead, the iron- and steel-making on the slide. The brass bands still play, though.

**West Yorkshire**, 'all wool and wuthering' – (a) the great jigsaw puzzle of the wool towns of Bradford, Halifax, Huddersfield, Leeds and their smaller satellite towns, now threaded by the M62, that grew rich and handsome on the work of the woollen textile mills, and (b) the Brontë Moors to the west: bleak heather moors where the Brontë sisters roamed and wrote romantic classics such as *Jane Eyre* and *Wuthering Heights*.

**North Yorkshire**, by far the biggest part – it sprawls over the top of the other three, and could swallow them all twice over. Here are Yorkshire's two National Parks: (a) the wide and lonely **North York Moors** in the north-east, bounded by North Yorkshire's fossil-rich 'Dinosaur Coast', and (b) the **Yorkshire Dales** in the middle and west, everyone's image of the North Country of TV series such as *Last of the Summer Wine* and *Emmerdale*. Their stone-built villages, green dales (valleys) and hillsides striped with stone walls are all jointed together by the knobbly spine of the **Pennine Hills** – the 'Backbone of Britain'.

## North-West

(Merseyside, Greater Manchester, Lancashire, Cumbria)

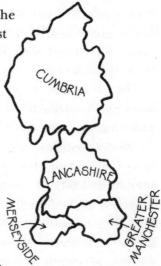

The north-west is probably the region of England with the least recognisable profile – and that's despite possessing the most famous beauty spot in the north. 'It's just . . . the M6 and the Lake District, isn't it?' No, not exactly.

The Beatles were north-westerners – they came from **Liverpool**, the capital city of **Merseyside**. Follow the M62 east from Merseyside for a few miles and you'll hit **Manchester**, Liverpool's fierce rival. For south-erners venturing north-west, these two great cities are the gateway to the region. Get beyond them and you reach Lancashire, once famous for cotton-spinning, now all but forgotten. Football fans – heard of Bolton, Blackburn, Burnley, Preston, Wigan, Bury, Oldham, Accrington? All Lancashire towns, all within a radius of thirty miles, many now gobbled up in Greater Manchester. Between them, and especially north of them, the wide and wild **Lancashire moors**, with the huge wheel of empty upland called the **Forest of Bowland** – wonderful hiking country – at their heart. Out to the west, the flat Lancashire coast of vast sandy beaches and wide muddy estuaries.

So up the M6 into **Cumbria**, and at Kendal or Penrith take a left turn into the **Lake District** National Park. But I

thought the Lake District was level with Scotland! Well ...
a surprising number of people think that, because of its
beautiful fells, or mini-mountains. But in fact, it's fifty miles
from Kendal to the nearest bit of Scotland; and, if you drove
due east from Kendal across England, you'd hit the coast of
North Yorkshire rather than anywhere within a hundred
miles of Edinburgh.

The Lake District is famous for its writers – poet William
Wordsworth, children's tale-spinner Beatrix Potter, writer of
walking guidebooks Alfred Wainwright – and, of course, for
the spectacular landscape that inspired them all, the tumbled
fells of granite and sandstone (including England's highest
peak, **Scafell Pike**, at 3,209 ft/978 m), and the long, slender
lakes. Not to mention the summer crowds, but that's another
story. To escape those, go to the Cumbrian coast for red
sandstone cliffs and unfrequented beaches. And Sellafield
nuclear reprocessing plant – but that, too, is another tale.

## North-East
(Durham, Tyne & Wear, Northumberland)

Oh, aye – it's grim up north, especially
up north-east. One long line of slag
heaps and shipyards, eh? Hmmm ...
it never was like that, even at the
height of heavy industrialisation, and
it certainly isn't like that now. The
eastern part of **County Durham** used to
be peppered with coal mines, but the last
of them closed in the mid-1990s; while
**Northumberland**, sitting above Durham,
lost its last mine in 2005. As for shipyards: Durham's great

ship-building town of Sunderland, at the mouth of the River Wear, saw its last yard close in 1988, and in 2006 the last Northumberland yard, in **Newcastle-upon-Tyne**, some twenty miles to the north, followed suit.

The death of coal-mining and ship-building has been an economic and cultural rabbit punch for the region. The worst of it has been in the area (some call it a county) of **Tyne & Wear**, which takes in the mouths of the River Wear with Sunderland and the River Tyne with Newcastle. Travel up the A1, though, and you'll see little of this. Newcastle has reinvented itself as Party Town. Former collieries have been flattened, steelyards and ironworks abolished in favour of warehousing, storage and lots and lots of superstores.

Turn off the A1, and surprise yourself. Explore west Durham and you'll find the beautiful **Weardale** and **Teesdale** moors; venture north from Newcastle up the Northumberland coast and discover fantastic empty beaches and a tooth-chattering sea. Up near the Scottish border, **Northumberland National Park** offers lonely moors, giant forests and the smoothly curving **Cheviot Hills**. Spanning the slender neck of England, the greatest Roman monument in Britain, Hadrian's Wall, marches the crests of the hills westward from Newcastle to Carlisle.

# 3

# LAND AND BORDERS – SCOTLAND

Y ou can split Scotland neatly enough into three regions, each one stretching like a band across the country, one on top of the other – **The Borders** at the bottom (south), **The Lowlands** in the middle, and **The Highlands** at the top (north). As for the counties within each region – there are so many, in such a tangle from too many reorganisations (what price Clackmannanshire, Kincardineshire and the Unitary Authority of Comhairle nan Eilean Siar, 'unitary in nature but not in name'?), that it's easier to go by the more recently created 'Council Areas' or local government regions of Scotland.

## The Borders

(Dumfries & Galloway, South Ayrshire, Scottish Borders)

Once you've gone north across the border between England and Scotland, you are (surprise! surprise!) in the Scottish Borders. It's odd that there's no corresponding area called the 'English Borders' – but there it is. This area has an amazingly bloody history. Three centuries ago, England and

Scotland were joined in the Act of Union (1707), but up till then the stretch of land along the border was known as 'The Debatable Lands', because both Scotland and England claimed it and fought continually over it, but neither seemed to own it. Instead, local raiding and feuding families ran it themselves, with 'might-is-right' as the only law, and murder and mayhem a fact of everyday life. And it still feels like a wild and woolly region.

Three Council Areas make up the Borders region. You cross the border into the biggest, **Dumfries & Galloway**, half a dozen miles north of Carlisle, as the English M6 turns into the Scottish A74 (M) motorway at Gretna Green (famous for its blacksmiths who used to marry runaway English couples in the forge). D&G fills the south-west corner of Scotland, and is big and sprawling. Very few travellers turn left at Carlisle to explore it, but those who do are in for a treat. There's a coast full of sandy bays, with some charming old-style towns done up by their arty communities – Kircudbright and Gatehouse of Fleet especially. Further west is the 'secondhand book capital' of Scotland, Wigtown, the gateway to the Isle of Whithorn peninsula where that crazy hippy legend *The Wicker Man* was filmed in 1972 – you can find the monster's leg stumps still rooted on the cliffs where Edward Woodward went out in a blaze of glory. From here D&G tails out in the big hammerhead peninsula of the **Rhinns of Galloway**, carved out of the sandstone by the Atlantic waves. Inland, D&G is a mess of wild, remote hills known as the **Southern Uplands**, many of them smothered under dark legions of conifer forests, legacy of a long-outdated, monoculture planting policy. **Galloway Forest Park** has lots of paths, bike trails and bridle tracks for exploring these lonely hills, and the **Southern Uplands Way** long-distance walking trail crosses them.

North of D&G sits **South Ayrshire**, a long, north-trending coast of dark red sandstone cliffs and beaches looking out to the huge volcanic lump of **Ailsa Craig**, way out to sea. This is Robbie Burns Country, based on the faded old resort of Ayr, where Scotland's roguish and passionate national poet was born (in Alloway), worked as a ploughman, drank, randied and wrote some sublime songs and poems.

If you head into Scotland up the A1 from the north-east of England, rather than up the M6 from Carlisle, you'll enter the third of the Borders Council Areas – this one, rather confusingly, with the name '**Scottish Borders**' all to itself. Not quite as big as D&G, its east coast is all craggy cliffs and sea-bitten bays, looking out on the North Sea. The interior is most people's dream of the Scottish Borders – neat little towns such as Galashiels, Peebles, Hawick and Selkirk, set in hill country that rolls like a green and purple sea from the rounded Cheviot Hills along the border with Northumberland. Beautiful long river valleys lie between the hills – Lauderdale, Teviotdale, Tweeddale – while the heights of **Ettrick Forest**, towards the Dumfries & Galloway border in the west, call to hikers and lovers of wild places.

## The Lowlands
(Central, Argyll & Bute, Stirling, Fife, Perth & Kinross, Angus, Moray, Aberdeenshire)

Parts of The Lowlands are almost as high as The Highlands, and plenty of Highland territory is lower than Lowland terrain. But let's not get into that. The Lowlands, for our purposes, slant north-east in a thick band right across the middle of Scotland.

**Central** is the first Council Area you hit as you travel north out of the Borders, and it comes as quite a shock if you were expecting the mountains and red deer to start straight away. Central stretches from coast to coast across the slimmest bit of Scotland, and it's the most populous, urbanised and industrialised sector of the country, with **Glasgow** on its west side and **Edinburgh**, the Scottish capital city, on the east. You're scarcely going to avoid this pair: industrial Glasgow and political, historical Edinburgh are the two most important cities in Scotland. They're only forty miles apart, connected by the M8 motorway. Each has its own airport and serves as the gateway to everything that's further north – Edinburgh to eastern Scotland, Glasgow to the west coast. The twin cities sit on a belt of coal and iron ore, much like the industrial regions of England, and each has its deep-water firth (Scottish for 'estuary') for ship-building and communication with the sea and the wider world – Glasgow the west-going **Firth of Clyde**, Edinburgh the east-facing **Firth of Forth**. So it's not surprising to find each city – but Glasgow in particular – surrounded by a whole jumble of ex-industrial towns (Motherwell, East Kilbride, Paisley, Milngavie) that are in search of a new identity.

Once you're past Central, you're into something more like Scotland – at least, the poster version. To the north and west of Glasgow is the Council Area of **Argyll & Bute**, a jagged-edged scatter of peninsulas and islands poking out and down towards Northern Ireland. The longest peninsula is Kintyre, 50 miles (80 km) from its northern clasp of the mainland to its southern tip near Paul McCartney's farm at the **Mull of Kintyre**. Out off the peninsula lie the neighbouring isles of Islay and Jura. Inland, the A82 'Road to the Isles' runs north-west across the vast bleak wastes of **Rannoch Moor** to enter the Highlands through the mountainous **Pass of Glencoe** – a dark, ominous landscape tainted by the notorious 1692 Glencoe Massacre of Macdonald clansmen here. The eastern part of Argyll & Bute holds the long and beautiful **Loch Lomond**, centrepiece of Scotland's first National Park and an absolute icon of Scottishness thanks to the traditional song 'The Bonnie Banks o' Loch Lomond' (*O ye'll tak' the high road and I'll tak' the low road, And I'll be in Scotland afore ye,* etc.). Loch Lomond was always a great destination for a day out from Glasgow, and so was the hill range of **The Trossachs**, just to the east and across the border in the Council Area of **Stirling**. This is Rob Roy Country, full of long thin lochs reflecting the hillsides where the jolly outlaw was supposed to have roved.

North of Edinburgh the Firth of Forth is spanned by two mighty bridges side by side, the famous Forth Railway Bridge with its three red dinosaur-spine spans, and the Forth Road Bridge. **Fife** stretches north-east from here, a peninsula of low farmland sandwiched between the firths of Forth (below) and Tay (above). Out at its eastern tip the tiny city of St Andrew's boasts a celebrated university and the world's best-known golf club, The Royal & Ancient.

Driving north up the M90 from the Firth of Forth you enter the big Council Area of **Perth & Kinross**, mountainous country where the 'backbone of Scotland', the Grampian Range, lumps up the landscape as it curves massively north-east. The A9 highway is the anchoring point here, linking Lowlands to Highlands as it forges north up Strathtay and Glengarry, with the Rannoch moors and peaks to the west and the mountainous Forest of Atholl to the north and east. East of Perth & Kinross the tumbled **Glens of Angus** run down to the fantastically weather-eroded red sandstone cliffs of the **Angus** coast. In this lower land lies a string of 'football towns' that any lover of BBC Radio's 'Sports Report' will instantly recognise as Scottish Division Two – Forfar, Arbroath, Brechin City, Montrose. 'East Fife Four, Forfar Five' – we're still waiting for that Golden Moment . . .

North of Angus run the final two Council Areas of the Lowlands: triangular **Moray**, its base facing north along the giant open mouth of the **Moray Firth**; and diamond-shaped **Aberdeenshire**, its broad coastal base pointing out into the North Sea like a blunt elbow around the fishing towns of Fraserburgh and Peterhead, its diamond-shaped apex burrowing into the Grampians at their highest and wildest around the **Cairngorms National Park**. Three major factors shape Aberdeenshire: (a) its elbow of a coast studded with fishing harbours; (b) the spectacularly beautiful strath (a broad river valley, as opposed to the narrow stream valley of a glen) of **Royal Deeside**, so called because of the Royal Family's residence there at Balmoral Castle; (c) the often stormy, often snowy **Cairngorm Mountains**, the nearest Britain possesses to an arctic plateau – a magnet for climbers, hill-hikers and skiers, and a fitting gateway to the Highlands.

## The Highlands
(Highland)

**Highland**, appropriately enough, is the only Council Area in the Highlands of Scotland.

The traditional gateway to the Highlands is the **Great Glen**, a geological fault in which rock faces slipped and tore a diagonal, dead-straight gash 6o miles (96 km) long across the face of Scotland, south-west to north-east. But Highland Council Area bulges east a bit, to take in the very high ground of the Grampian mountain range between the Great Glen and the Cairngorm Mountains. If you're coming from Edinburgh up the A9, you'll arrive in **Inverness**, 'capital' of the Highlands, at the top (north-east) of the Great Glen. If the A82 from Glasgow is your route, you'll be in the Highlands much sooner, because the A82 enters the region at Fort William at the south-west end of the Great Glen, very near the highest mountain in the British Isles, **Ben Nevis** (4,409 ft/1,344 m/13 Big Bens).

The Great Glen itself holds a string of long, narrow lochs, joined up by the **Caledonian Canal** – the best known is **Loch Ness** near Inverness, where eyesight and imagination both need to be on the powerful side to spot Nessie, the Loch Ness Monster.

Once across the Great Glen, you are in the Highlands proper. Here in the north-west and extreme north of Scotland you'll find the real rough stuff – the huge herds of red deer, the lonely coasts of sea otters and golden eagles, the isolated mountain ranges where few venture, the giant bogs, the empty glens and abandoned villages. The Highlands comprise the wildest and emptiest land in Britain. The basic rock is some of the most ancient in these islands – hard white quartzite, schist and sandstone, running out west into what *is* the oldest rock in Britain: a speckled granite called gneiss, up to 3,300 million years old – about a quarter of the way back to the Big Bang. Roads in the north and west Highlands are few and long and winding, towns scarce, cities nonexistent. Mountains, glens and lochs are plentiful. Natural beauty and grandeur are everywhere you look. Scotland's native language, Gaelic, and the traditional culture that goes with it are strongest here.

The whole west coast is cut into remote peninsulas. Within the first 50 miles (80 km) as the cormorant flies (but three times that by road), you find the ferry ports of Oban, Mallaig and Kyle of Lochalsh, each serving a different cluster of islands, each embedded in its own bay. North of Kyle it's even wilder and more remote, a 100-mile (160-km) arc of peninsulas, sea lochs and narrows up to **Cape Wrath** (splendidly named), the stormy north-west corner of Scotland. Inland, the Highlands are made up of a series of east–west glens and straths, each hemmed in on either side by sloping walls of mountains. Roads wriggle and snake to get where they're going. This is a majestic landscape, rich in wildlife and grand scenery, but stripped of its human inhabitants – most were evicted to make way for more profitable sheep during the notorious Highland Clearances of the nineteenth century.

Up towards the north coast, and the further east you go, the landscape begins to flatten and roll with a lowland feel. The mountains stand out individually, dramatic features in the vast flat boglands of the **Flow Country**, a million acres of blanket bog threaded by a couple of bumpy roads and a single railway line. Scotland's north coast is dark, bleak, craggy and wind-whipped. You could follow the A99 as far as John O'Groats, if you really felt the pull of the name, but it's pretty disappointing when you get there. Better to go on out a mile or so east to Duncansby Head, the real north-easternmost point of mainland Britain, or a few miles west to Dunnet Head, the most northerly spot. Come fair weather or foul, you'll never forget the view of the tides in the Pentland Firth, nor the shadowy shapes of the **Orkney Isles** beckoning you on over the northern sea horizon.

# 4

## LAND AND BORDERS – WALES

Just like England and Scotland, the counties of Wales have had their share of renaming – from ancient kingdoms such as Gwent and Powys, through Anglicised styles (Merionethshire, Monmouthshire) and back to Welsh names again. And even then the bureaucrats haven't been content – viz. dear old Gwent, that's now supposed to be referred to as the 'county-bloody-boroughs' of Blaenau Gwent, Newport, Tolfaen, Bridgend, Caerphilly, Merthyr Tydfil, Rhondda Cynon Taff, Vale of Glamorgan and Neath Port Talbot. Unless it isn't.

Let's keep it simple, stupid, and stick to what are now known as the 'preserved counties of Wales' – ancient names for counties everyone can recognise.

### Gwent

'*Croeso i Cymru*' says the noticeboard by the M4 as you drive off the Severn Bridge and away from England. 'Welcome to Wales'. Even though you'd have to carry on for another hundred miles into West Wales to hear much Welsh spoken, it's a reminder, if you're English, that you're not at home now, and if you're Welsh that you're back in the land of your fathers. As a first taste of Wales, Gwent is green and pleasant, especially in its eastern part, the old county of Monmouth-shire near the English border, with its wooded **Wye Valley** and gentle hills. Further into Wales, though, the land begins to steepen, with the M4 running west past Newport at the mouth of deep river valleys, Ebbw and Rhymney, which twist away north. Turn right off the motorway and follow the snaky roads up these steepening valleys through Blackwood (home town of angst-rock band The Manic Street Preachers), Abertillery, Tredegar and Ebbw Vale, and you'll be entering the extraor-dinary post-industrial landscape of the Welsh Valleys.

## Mid-Glamorgan

Mid-Glamorgan contains everything that you picture when you think of the **Welsh Valleys** – the steep-sided clefts of Taff Vale and Rhondda, with bleak moorland hills islanded between them; rivers, roads, railways and canals all squeezing together through the long, thin towns in the valley bottoms; hillsides striped with endless lines of red-brick terraced houses. Every-thing is so quiet around here these days that the history is hard to credit, but the fact is that the Welsh Valleys were the power-house of the world in the mid-nineteenth century. They sit on incredibly deep and rich seams of coal, perfect for producing steam, along with enough iron ore to keep the furnaces of the Industrial Revolution busy day and night. And they wind southwards into South Glamorgan, to the sea and the docks around Cardiff – the ideal outlet for the coal and the iron.

Turn up the **Taff Valley** through Pontypridd (home town of Tom Jones, lest we forget), and you'll come to Merthyr Tydfil at the top of the valley. Merthyr didn't even exist as a town in the mid-eighteenth century. A hundred years later it had become the greatest ironworking town in the world. The furnaces roared day and night. The yearly output from its Dowlais Ironworks alone could have built a thousand heavy steam locomotives, or laid a railway track from Cardiff to the north of Scotland and back. Now there's no ironworking in Merthyr. Hop over the next valley to the west, Cwm Cynon (home of gruff-rock band Stereophonics), and the valley after that is the most famous in South Wales coal-mining history, **Cwm Rhondda**. Coal production came long after iron-making in the Welsh Valleys. When Dowlais was in its Victorian heyday over at Merthyr Tydfil, there were fewer than 3,000 people living in Cwm Rhondda. But half a century later, at the time of the First World War, 160,000 were crammed in here. A miner would get into a bed kept warm by the previous occupant, and would warm it himself, as he slept, for the next shift-worker. The mines around this one Welsh valley produced about one-twentieth of all the coal being burned in the houses, factories and steam-powered machines of Britain. Staggering figures that added up to hard work and poor health for the miners; riches for the mine-owners. Nowadays no coal comes out of Cwm Rhondda. Underinvestment, cheap imported coal and a turning away from pollutant power have killed the iron and coal industries. But the slag heaps, the mountains of mine spoil, still remain, grassed over like hills, only their geometric shape revealing their origin.

## South Glamorgan

South Glamorgan sits neatly to the south of Mid-Glamorgan, with the M4 separating them. South Glamorgan possesses only one city, but that's the most important one in Wales. **Cardiff** and its docks were the gateway to the world for the products of the industrial valleys, but these days the capital city of Wales is the seat of the Welsh Assembly, and a bit of a party town, with massive redevelopment of its docklands.

Out west of Cardiff and south of the M4 is the pastoral hinterland of the Vale of Glamorgan, bounded by the Bristol Channel, whose lack of through roads and cut-off position – thanks to the motorway – make it superb for walking and just idling about, especially along the beautiful **Glamorgan Heritage Coast**.

## West Glamorgan

West Glamorgan is directly west of Mid-Glamorgan, and shares the westernmost bit of the Welsh Valleys. But this county mostly consists of lovely forested hills inland, with the M4 running west along the coast until it reaches industrial territory again, unmistakably so when the smoking stacks of Port Talbot steel-works heave into view. Just beyond, a left turn brings you through **Swansea**, Wales's Second City – 'Copperopolis' to locals in the nineteenth century, when 90 per cent of the world's copper was smelted right here. Dylan Thomas, the Welsh national poet, was born in Swansea, and he grew to love the gorgeous green **Gower Peninsula**, which stretches south-west from the city into the Bristol Channel. The Gower is one of those magical places where time seems to slip its leash – especially out at the far end, exploring the dramatic tidal promontory of **Worm's Head**. Go out there, and you'll see what I mean.

## Dyfed

As Cornwall is to England, so the Dyfed coast is to Wales: seaside holiday country. Here the bucket and spade do more for the economy than the blast furnace and the coal shovel. Dyfed is the biggest Welsh county, and the most westerly. Down at the south-westernmost tip is the fractured, sea-eroded coastal county of Pembrokeshire, now preserved from overdevelopment as the **Pembrokeshire Coast National Park**. The Pembrokeshire Coast Path National Trail threads the cliffs and coves; there are long-established seaside resorts, a whole string of sandy beaches, and boat rides out to Skomer, Skokholm and the other islands. **St David's**, Britain's smallest city, sits out here, and at St David's Head the Dyfed coast swings north-east for a great 60-mile (about 100 km) run up the side of **Cardigan Bay**.

Inland, Dyfed ripples west from the Black Mountain north of Swansea to the **Presceli Hills** out in western Pembrokeshire. Further north, the landscape is dominated by the high, well-forested spine of the **Cambrian Mountains**, criss-crossed with ancient drovers' roads, that parallels the northward curve of Cardigan Bay. This is deep hill country, very wild and lonely...

## Powys

... as is Powys, the next-door neighbour to the east of Dyfed, a ragged-edged, roughly rectangular county 'descended' from an ancient British kingdom back in the mists of time. Powys has no coast – it's a huge region, almost as big as Dyfed, of landlocked back country, with the **Brecon Beacons** National Park on its southern doorstep, the **Cambrian Mountains** marching up its western flanks, and the Welsh/English border forming its entire eastern side. If you want to see the remote heartland of Powys – and of Wales, for that

matter – just follow the A470 north from the Brecon Beacons to Builth Wells, then the A483 the rest of the way up to Oswestry. It's a hundred-mile journey through a hilly landscape, with plenty of opportunity to sidetrack – for example, west to Fforest Fawr opposite the Brecon Beacons, east to climb the Brecon Beacons themselves and to explore the strange, boat-shaped **Black Mountains** beyond; west to the boggy wastes of **Mynydd Eppynt**, the **Elan Valley** with its string of jewel-like reservoir lakes, and – in the north of Powys – the haunting, misty **Berwyn Hills**, paradise to any determined walker intent on exploring wild country with GPS, map and compass. Here tumbles **Pistyll Rhaeadr**, the highest waterfall in Wales, symbol of all that is wild and mighty about Powys.

## Gwynedd

This is what people mean when they talk of the Welsh mountains. Here in the north-west corner of Wales rises **Mount Snowdon** (3,560 ft/1,085 m), the tallest mountain in Wales and the centrepiece of the **Snowdonia** National Park, at the edge of a ring of magnificent mountains, the **Carneddau** and the **Glyders**. These great structures of slate, gritstone and mudstone, topped off with volcanic caps, look and feel majestic, and climbers and walkers flock to them. But they're not the only things worth looking at in Gwynedd.

The outer mountains were burrowed by slate miners into vast ledges and caverns – truly weird scenery. Further south, the twin mountains of **Cadair Idris** and **Plynlimon** guard the southern borders of Gwynedd, just inland of a Cardigan Bay coast split by big sandy estuaries with seaside resorts – Aberdyfi on the Dyfi, Barmouth on the Mawddach, and Porthmadog on the estuary shared by the rivers Dwyryd and Glaslyn (where you'll find the fantasy village of **Portmeirion**, surreal setting for the freaky 1960s cult TV show *The Prisoner*). From here the long snout of the **Llŷn Peninsula** sticks way out west.

The northern coast of Gwynedd is dominated by the huge bulk of the **Isle of Anglesey**, separated from the mainland by the very narrow **Menai Strait** – it's only half a mile wide for most of its 15-mile (24-km) length. You'd cross the strait by one of two lovely old bridges, road or railway, if you were making for Holyhead and the Dublin ferry.

## Clwyd

As Dyfed and Pembrokeshire mean seaside holidays to folk from South Wales and the South Midlands, so **Clwyd** and the coast of North Wales ring the same bell around Birmingham and Liverpool. Here are the sandy resorts of Llandudno, Colwyn Bay, Rhyl and Prestatyn, strung along the coast, with castles frowning over charming ancient towns behind them – Ruthin, Denbigh, and Conwy with its medieval town walls. The whole of central Clwyd is boxed in by high country – the mountains of **Snowdonia** butt up against the western border of Clwyd, the **Clwydian Hills** themselves form a spine of knobby uplands towards the east, and in the south stand the **Berwyns**.

Over on the eastern side of the county, things become more gritty and industrial the nearer you get to the English border, with the heaps and holes of quarrying and iron-mining, not to mention a string of former ironworking towns in search of a new identity. Up on the coast here you look across the wide mudflats of the Dee Estuary to a huge wind farm and the dominant bulk of Shotton Steelworks (now making structural components under the Corus name), a curiously downbeat and pragmatic farewell to the land of poetry, song and romance.

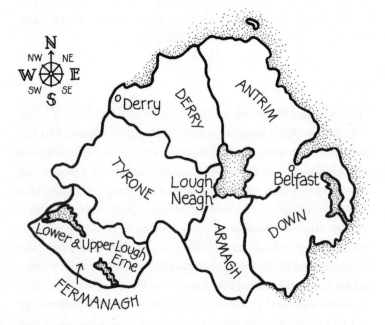

# 5

## LAND AND BORDERS – NORTHERN IRELAND

'Why go there? That place spells trouble.' You might have been forgiven for thinking that during most of the late twentieth century, and – even now – only a fool would say that everything is completely hunky-dory. But Northern Ireland has picked itself up after the Troubles, dusted itself off, and started . . . if not all over again, then at least from a better and more optimistic place. 'Undiscovered gem' is a cliché bashed to death by the tourist industry, but for the mainland Briton exploring these islands it's pretty much true of Northern Ireland.

Northern Ireland can be handily broken up into an anticlockwise tour of the six counties of Antrim, Derry,* Tyrone, Fermanagh, Armagh and Down, which cluster round Lough Neagh, the largest inland sheet of water in the British Isles. Antrim, Derry and Down have coastlines, all facing

---

* The town and county of Derry are known thus to those who look politically to a united Ireland, or as 'Londonderry' by those who support links with the United Kingdom. Nothing at all should be inferred from the writer's terminology – 'Derry' is simply the more historic name.

either north, east or south towards the British mainland; the other three counties are landlocked.

## County Antrim

Northern Ireland's capital city of **Belfast** sits sheltered in the crook of **Belfast Lough**. It's a town with a great manu-facturing heritage, including shipyards – *Titanic* was built here – and linen mills among other industries, and a tough recent past. That's all behind it now, and visitors are coming for the shopping, the nightlife and the strong black humour.

When visitors leave Belfast, it's mostly north up the A2, making for the Antrim coast. This is one of the greatest drives in the British Isles, past a succession of ever taller and ever more colourful cliffs. Basalt spilling from a volcanic explosion formed the cliffs when it solidified, and rain, frost and tumbling streams carved out the gorgeous **Glens of Antrim** running down to the cliffs. In glorious scenery the Antrim coast turns a right angle and runs west to the **Giant's Causeway**. Northern Ireland's No. 1 tourist attraction, a basalt ledge of hexagonal columns sticking out into the sea, can be very underwhelming when seen in dull weather among crowds of sightseers, but come when the wind's howling and heavy seas are crashing over it, and you'll get the point.

## County Derry

The basalt coast slips westward from County Antrim into County Derry, and the A2 slips with it. On the cliffs stands the eerie ruin of Downhill Palace, above beautiful sandy beaches and flowery dunes; here you turn inland through

Limavady to **Derry City**, snug and small behind its seventeenth-century walls in the throat of Lough Foyle. County Derry is rich farmland with plenty of old woodland, running south to the start of a rise of ground where the underlying quartzite and schist begin to push up into the rounded forms of the **Sperrin Hills**. Down in the farming country of the south-east, meanwhile, basalt underpins the boggy dairying country around Castledawson, the home ground of Northern Ireland's national poet and Nobel Laureate, Seamus Heaney.

## County Tyrone

Tyrone has an underdeveloped tourist honeypot in the shape of the **Sperrin Hills**. This triple range of dome-like uplands, cut east–west by the two deep valleys of the Glenelly and Owenkillew rivers, should be a walker's paradise – and will be when the number of walker-friendly paths across the Sperrins is increased. The town of Omagh towards the west of the county has the Ulster-American Folk Park, a great outdoor reconstruction museum, on its doorstep; and halfway along the road from Omagh to Cookstown you'll find An Creggan centre in the midst of some remarkable ancient sites – stone tombs and stone circles from the deep past, preserved thanks to the thick bog that developed here thousands of years ago.

## County Fermanagh

Down in the south-west of Northern Ireland, County Fermanagh is all about limestone and lakes. One-third of Northern Ireland's 'Lake District' is under water – the great double lake system of **Lough Erne**. Upper Lough Erne is fractured into hundreds of small lakes and lakelets, wonderful for fishing and casual boating; Lower Lough Erne to the north is one huge sheet of water, superb for sailing, cruising and visiting the islands with their remarkable monastic remains and unique ancient stone carvings. Pinched between the upper and lower lough lies the county capital, Enniskillen, and from here you travel south towards the border with the Republic of Ireland, through limestone country. Rain and rivers have eaten the limestone into spectacular caverns; at **Marble Arch Caves** you can boat through this memorable underworld.

## County Armagh

'Bandit Country! Don't go there!' County Armagh got that reputation during the Troubles, when there seemed nothing but bad news from its **South Armagh** area close to the border with the Irish Republic. But you'll regret it if you don't go down and wander through the lumpy landscape there, a tightly drawn and intimate dairy-farming landscape of small steep hills. These are drumlins, heaps of gravel left behind when the melting glaciers retreated at the end of the last Ice Age. The gentle influence of rain and wind has gradually reduced and rounded them, as it has the far more abrupt profile of the **Ring of Gullion**, a circle of stumpy basalt mountains 10 miles (16 km) across, with the volcanic ash heap of **Slieve Gullion** mountain rising higher than all at the centre. Further north you pass through apple orchards and the ancient city of **Armagh** with its two splendid cathedrals

– one Protestant, one Catholic, naturally – before reaching the M1 motorway and the shores of **Lough Neagh**. This is a huge sheet of water. Using the popular CRCS (County of Rutland Comparison Scale), the lough is – in fact – the same size as Rutland, but more full of eels and fossilised wood.

## County Down

County Down has the most indented coast of the six counties, with the **Ards Peninsula** projecting outward and downwards from Belfast for nearly 20 miles (32 km) – the upper beak of the 'parrot' as it defies the hag across the sea. The Ards shelters the bird-haunted tidal inlet of **Strangford Lough**, one of the most atmospheric places in Britain, with its vast mudflats, huge skies and string of islands. Further south the **Lecale Peninsula** bulges east in a rounded coast of cliffs and tiny fishing villages; then comes the pride of Down, the granite peaks of the **Mourne Mountains**. These are very fine mountains, crossed with paths, rising to the shapely peak of **Slieve Donard** (2,795 ft/852 m). The Mourne summits are linked by the famed Mourne Wall, built in the hungry years of the early twentieth century to provide wages for unemployed and desperate men and their families. So history, bitter and indelible, can be scarred into a landscape.

# 6

## TWENTY-ONE CITIES

The cities of Britain didn't just pop up like mushrooms. Their geography made them what they are, gave them a reason to exist, offered them a story, and had a hand in their tale of industrial rise, their post-industrial decline and fall, and their long road to recovery. These places have taken a terrible battering, most of them, from wartime bombs and industrial slump and political ideology and bad planning. We see them every time we travel the motorways or the railways, or look down from an aeroplane. We notice their changes, like the faces of friends not seen for years. 'Oh, God! What's *that?* What have they *done* to — [insert city of choice]? Where's — [insert landmark of choice] gone? God, how *could* they . . . ??'

But cities don't give up. Their football crowds chant their praise in scurrilous, scabrous scatologies. They mount festivals, gay pride marches, battles of the bands. They make cruel jokes about rival cities. The spirit of Bristol and Glasgow, Swansea and Belfast is something to admire, something to celebrate. Their stories are ours, for better or worse, after all. Here are twenty-one cities that fill in the jigsaw of our national character. For all the others, the forgotten places

at the end of a line to nothing, consult www.knowhere.co.uk
– a very funny, very rude guide to small-town Britain.

# ENGLAND

## West Country

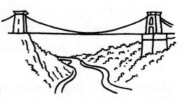

*Bristol*
*population*: 420,000.
*football teams*: City ('Robins')
at Ashton Gate; Rovers ('Gasheads') at Memorial Stadium.
*bands*: Massive Attack, Adge Cutler and the Wurzels,
Portishead.
*famous folk*: Banksy (1974– ), guerrilla artist; Cary Grant
(1904–86), actor; Damien Hirst (1965– ), shock artist; J.K.
Rowling (1965– ), *Harry Potter* author; Michael Redgrave
(1908–85), actor/director; Robert Southey (1774–1843),
Poet Laureate; W.G. Grace (1848–1915), cricketer.
*famous joke*: Q: How do you kill a City/Rovers fan while he's
drinking?
A: Slam the seat down on his head.
*matchbox history*: A famous seaport, but the sea's actually
4 miles (6 km) away at the end of the Avon Gorge, which
is winding, tidal and steep-sided. That didn't matter when
ships were small, so Bristol – facing west towards Britain's
colonial possessions across the Atlantic – became hugely
rich through trading tobacco, sugar and slaves. Then
vessels became too big to negotiate the gorge, new docks
were built at its mouth, and Bristol lost its ships.

The great Victorian engineer Isambard Kingdom

Brunel, fabulously named, left the city a fabulous legacy, including the world's first propeller-driven ocean liner, SS *Great Britain*, the Great Western Railway, the Clifton Suspension Bridge across Avon Gorge, and a properly clean 'Floating Harbour'.

*there and back*: Road: M4 (London and South Wales); M5 (Exeter and Birmingham).

Rail: Bristol Temple Meads (London and Wales), Bristol Parkway (Midlands).

*go there for*: great music; Ashton Court park; Avon Gorge rock climbing; walks or water taxi round the Floating Harbour; I.K. Brunel's engineering masterpieces (SS *Great Britain*, Clifton Suspension Bridge); brightly painted terraced houses perched on hills; hot-air balloons.

*festivals*: Bristol Harbour Festival (July/August: water-based); Bristol International Balloon Fiesta (August); Bristol Festival (September: music, arts, circus).

*mad fact*: Sarah Ann Henley, aged twenty-two, jumped from Clifton Suspension Bridge on 8 May 1885 after a lover's tiff, fell 250 ft (76 m) to the River Avon, and survived – her skirts blew out and acted like a parachute, landing her gently in the mud. She lived till she was eighty-five.

*on the way to*: Glastonbury Festival; West Country seaside holidays; cider heaven.

*psychic smell*: earthy.

## South Country

*Portsmouth*

*population*: 200,000.

*football team*: Portsmouth FC ('Pompey') at Fratton Park.

*bands*: The Cranes.

*famous folk*: Isambard Kingdom Brunel (1806–59), engineer; Jeremiah Chubb (1793–1870), inventor of the Chubb lock; Charles Dickens (1812–70), novelist; Paul Jones (1942– ), rock singer; Peter Sellers (1925–80), actor/comedian.

*famous joke*: Southampton fan 1: 'Two Pompey fans jump off a cliff – who hits the ground first?'

Southampton fan 2: 'Who cares?'

*matchbox history*: With its islands, peninsulas and deep water inshore, and its position guarding England's south coast against attack from France, Portsmouth has been a harbour and port since Roman times, a naval base for 800 years, and the Royal Navy's 'home port' since the time of King Henry VIII. Admiral Horatio Nelson left Portsmouth on 14 September 1805, his final steps on English soil before his death at the Battle of Trafalgar a week later. The Royal Navy still maintains a dockyard and base here, next to the Historic Dockyard where Nelson's flagship HMS *Victory*, the remains of Henry VIII's flagship *Mary Rose* and other splendid ships are preserved. Newest attraction: the eye-catching, 558-ft (170 m) tall Spinnaker Tower, right on the waterfront.

*there and back*:   Road: M27 (Southampton); A27 (Brighton); A3 (London).

Rail: Portsmouth Harbour.

*go there for*: the view from the Spinnaker Tower; ships and museums in the Historic Dockyard; D-Day and Charles Dickens Birthplace Museums; shopping at Gunwharf Quays.

*festivals*: International Kite Festival (August); Maritime Folk Festival (September); International Festival of the Sea (every few years).

*mad fact*: In 1999 Portsmouth FC were so hard up they cancelled their weekly order of clean jockstraps for the team.

*on the way to*: Isle of Wight; D-Day tapestry at Southsea.

*psychic smell*: rumbustious.


## Home Counties

*London*

*population*: 7,500,000.

*football team*: Chelsea FC ('The Blues') at Stamford Bridge – and some others.

*bands*: The Rolling Stones, The Who, Madness, The Sex Pistols, The Clash, The Kinks, The Libertines, The Rakes . . .

*famous folk*: Alan Sugar (1947– ), businessman; Alfred Hitchcock (1899–1980), film-maker; Lady Violet Bonham Carter (1887–1969), politician; Ronnie Kray (1933–95), gangster; Henry VIII (1491–1547), monarch; Henry Cooper (1934– ), pugilist; HM Queen Elizabeth II (1926– ), monarch . . . and many, many more.

*famous joke*: City banker: 'Constable, someone's stolen my Porsche! I tried to grab the door handle, but he drove away!'

Policeman: 'Sir – he's torn your whole arm off!'

City banker: 'Oh, my God! . . . My Rolex!!!'

*matchbox history*: The Romans founded 'Londinium' near the mouth of the River Thames because that was the first place going inland that they could put a bridge across. With good sea communications, proximity to the Continent, a mild climate and fertile soil, London could only prosper – and it did. Roman roads radiated from the city:

Stane Street to the south coast, Watling Street north-west from Dover to Chester, the Portway south-west into the West Country, Akeman Street west to Gloucester and Ermine Street all the way north to York. And the routes they opened up still form the strands of England's road system today, with London the fat spider at the hub of the web. In the heyday of the British Empire, London was the most important city in the world; it's still the magnet that draws every visitor to these shores, and a mazy mix of immigrants.

*there and back*: Road: M25 (circular), M4 (west), M3 (south-west), M23 (south), M20 and M2 (south-east), M11 (north-east), A1M (north), M1 and M40 (north-west). Rail: many mainline stations.

*go there for*: theatre, art galleries, heritage, history, a world of international food and culture, nightlife, daydreams, music, dance and (yes, unfortunately they've arrived) Dunkin' Donuts.

*festivals*: Baishakhi Mela/Bengali New Year (May); Trooping the Colour (June); City of London Festival (June/July); Pride London (June/July); The Proms (summer); Notting Hill Carnival (August); Underage Festival (August); Diwali on the Square (October); Lord Mayor's Show (November); Carols by Candlelight (December) ... and many more.

*mad fact*: In 1945 Big Ben's clock lost five minutes when a flock of starlings landed on the minute hand and sat there.

*on the way to*: the rest of England if you're coming from Europe – or vice versa.

*psychic smell*: imperial.

## East Anglia

### *Cambridge*

*population*: 120,000.

*football team*: Cambridge City
    FC ('The Lilywhites') at the
    City Ground.

*bands*: Pink Floyd.

*famous folk*: David Gilmour
    (1946– ), guitarist; John
    Maynard Keynes (1883–1946), economist; Eddy Shah
    (1944– ), newspaperman; G.A. Henty (1832–1902),
    historical novelist; Ronald Searle (1920– ), cartoonist;
    Judith Weir (1954– ), composer.

*famous joke*: Aussie in Cambridge University library: 'G'day,
    mate – which shelf's the Descartes on?'

    Undergraduate: 'I say, you shouldn't end a sentence with
    a preposition.'

    Aussie: 'Oh, okay – which shelf's the Descartes on, ya
    stupid bloody Pom?'

*matchbox history*: Around 800 years ago a group of lay scholars
    who'd been expelled from Oxford for rowdy behaviour
    ended up here in the misty Fens on the River Cam, where
    some monastic houses of learning already existed. At first
    the students lived scattered about, but in 1284 the first
    college, Peterhouse, was founded to get them all under
    one roof for better convenience and discipline – they
    were a wild lot. The colleges multiplied, and developed
    a reputation, along with Oxford, of being the world's
    finest university. The Fenland town that grew around the
    university has its own fine reputation for technological
    industries, and of course for the colleges' sensationally
    beautiful architecture.

*there and back*: Road: M11 (London), A14 (Midlands and Felixstowe).

Rail: Cambridge.

*go there for*: a superior education; stroll around the colleges; punting along The Backs lawns; picnic in Grantchester meadows.

*festivals*: Cambridge Winter Ale Festival (January); Midsummer Fair (June); Cambridge Folk Festival (July/August); Cambridge Festival of Ideas (October).

*mad fact*: Immodest and/or incontinent women are not permitted to dwell within four miles of Cambridge.

*on the way to*: Wicken Fen National Nature Reserve; Newmarket Races; North Norfolk coast; Ely Cathedral.

*psychic smell*: misty.


## South Midlands

*Birmingham*

*population*: 1,000,000.

*football teams*: Birmingham City ('The Blues') at St Andrew's Stadium;
Aston Villa FC ('The Villa, The Villans') at Villa Park.

*bands*: The Move, The Moody Blues, Black Sabbath, Duran Duran, UB40, Editors, Cradle of Filth.

*famous folk*: Noddy Holder (1946– ), Slade singer; George Cadbury (1839–1922), chocolate-maker; Francis Galton (1822–1911), meteorologist and scientist; Jasper Carrot (1945– ), comedian; Abraham Darby (1677–1717), cast-iron and blast-furnace pioneer; Lenny Henry (1958– ), comedian and actor.

*famous joke*: Stranger: 'Scuse me, mate, d'you know if there's a B&Q in Birmingham?'

Brummie: 'Don't know, mate – just a minute, let's see . . . B-I-R-M-I-N-G-H-A-M . . . ummm . . . don't think so, mate.'

*matchbox history*: Birmingham sits squarely on coal and iron ore, the two items needed for iron-making. So from medieval times the little market town in the centre of England was surrounded by iron-making villages. With the advent of canals, and then railways, to bring raw materials and take away finished goods, the conurbation swelled into a powerhouse of the Industrial Revolution. Dozens of villages on the north-west outskirts joined up to become the Black Country; heavy industry went on there (chain-making, iron-forging), while in Birmingham flourished the craftsmen (jewellers, gun-makers). Recently the Black Country has struggled with post-industrial dereliction – but the dark humour and 'can-do' attitude of the region are still very much to the fore.

*there and back*: Road: M42 and M6 Toll (circular – partly!), M6 (M1 and the north-west), M5 (Bristol).
Rail: Birmingham New Street, Birmingham NEC (National Exhibition Centre).

*go there for*: National Exhibition Centre; Bullring shopping centre; great Indian food in the 'Balti Triangle'; canal boating and walking.

*festivals*: National Boat, Caravan and Outdoor Show (NEC, February); Crufts Dog Show (NEC, March); International Dance Festival (April/May); International Jazz Festival (July); Birmingham Carnival (August); National Wedding Show (NEC, October); International Motorcycle and Scooter Show, and Good Food Show (both NEC, October).

*mad fact*: The oft-heard boast, 'Birmingham has more canals than Venice,' is true. Venice's 26 miles (42 km) is trounced by Brum's ... 35 (56 km)!

*on the way to*: Cannock Chase's leafy open country (north); Shakespeare Country (south); the Wyre Forest (west); Coventry's remarkable modernist cathedral (east).

*psychic smell*: loud and proud.

### Oxford

*population*: 150,000.

*football team*: Oxford United FC ('The U's, the Yellows, The Boys from Up the Hill') at the Kassam Stadium.

*bands*: Radiohead.

*famous folk*: Basil Blackwell (1889–1984), bookseller; Richard the Lionheart (1157–99), monarch; Tim Henman (1974– ), tennis player; Jacqueline du Pré (1945–87), cellist; Stephen Hawking (1942– ), physicist and author; Orlando Gibbons (1583–1625), composer; Dorothy L. Sayers (1893–1957), novelist ... and lots more.

*famous joke*: Q: How many Oxford undergraduates does it take to change a light bulb?

A: None – Daddy will buy them a new house.

*matchbox history*: Founded about forty years before Cambridge, also by students who had been ejected – this lot from Paris University. King Henry II offered them the protection of one of his royal residences on the River Thames, and the colleges of Oxford developed here. Sisters and rivals, Oxford and Cambridge became the world's finest universities. Publishing and printing were natural industries for Oxford, and car-making became a big employer in a city sitting squarely on the route from London to the Midlands, with good road, river, canal and railway connection. That's why you go through a thick shell of industrial plant and housing to reach the beautiful, historic old core of Oxford.

*there and back*: Road: M40 (London and the Midlands), A40 (London and Cheltenham).

Rail: Oxford.

*go there for*: learning and civilisation; fabulous architecture; black gowns and bicycles (catch 'em while you can).

*festivals*: Oxford Literary Festival (March); Oxford May Music Festival (May); Elder Stubbs Festival of allotments, vegetables and more (August); Oxford Chamber Music Festival (September/October).

*mad factoid*: Ex-US President Bill Clinton smoked cannabis in University College . . . but didn't inhale.

*on the way to*: Blenheim Palace; east Cotswold Hills; Ot Moor wetland nature reserve.

*psychic smell*: brainy.


## North Midlands

*Stoke-on-Trent*
*population*: 240,000.
*football teams*: Stoke City ('The Potters') at the Britannia Stadium; Port Vale FC ('The Valiants, The Vale') at Vale Park.
*bands*: Motörhead.
*famous folk*: Arnold Bennett (1867–1931), Potteries novelist;

Josiah Spode (1733–97) and Josiah Wedgwood (1730–95), master potters; Robbie Williams (1974– ), singer; Stanley Matthews (1915–2000), footballer; Ian Kilmister a.k.a. Lemmy (1945– ), bass monster.

*famous joke*: Wedgwood to potter: 'What are you doing smoking and drinking and playing cards outside that oven?'

Potter: 'Just kil'n time, sir.'

*matchbox history*: Under the Six Towns of Stoke-on-Trent – Tunstall, Burslem, Hanley, Stoke, Fenton and Longton – lie coal, iron and lots of potter's clay. Once the pottery industry got mechanised, the towns ran together into a smoking, flaming, roaring, polluted industrial giant. Master potter Josiah Wedgwood organised the workers and built canals to take the breakable products safely away. Messrs Spode, Minton, Doulton and other master potters became hugely rich, their ware world-famous, their town the most dug-up, wrecked and fouled in Europe. Now that the industry's all but dead – thanks to foreign competition – Stoke has had a huge clean-up, making clay holes into stadia, spoil heaps into hills and railway tracks into paths and cycleways. The potteries themselves? Long gone under housing and light industry.

*there and back*: Road: M6 (Birmingham and Liverpool), A50 (M1 near Derby, Nottingham).

Rail: Stoke-on-Trent.

*go there for*: pottery museums; parks and cycleways; canal walking.

*festivals*: Stoke-on-Trent Music Festival (all year).

*mad fact*: At the height of the Potteries industry, a butcher pastured some cows on the town moor overnight. When he slaughtered them next day, their innards had turned completely black.

*on the way to*: Staffordshire Peak District.

*psychic smell*: reborn.


## Yorkshire

*Sheffield*

*population*: 530,000.

*football teams*: Sheffield United ('The Blades') at Bramall Lane; Sheffield Wednesday ('The Owls') at Hillsborough Stadium.

*bands*: Def Leppard, Arctic Monkeys, ABC, The Human League, Pulp, Joe Cocker.

*famous folk*: David Blunkett (1947– ), politician; A.S. Byatt (1936– ), novelist; Peter Stringfellow (1940– ), nightclub owner; Joseph Locke (1805–60), engineer; Prince Naseem Hamed (1974– ), pugilist (not really a prince); Harry Brearley (1871–1948), inventor of stainless steel.

*famous joke*: Owls fan 1: 'What's the difference between the Blades' defence and a taxi?'

Owls fan 2: 'A taxi won't let six in.'

*matchbox history*: By Shakespeare's day, Sheffield was already Britain's centre of cutlery production. The town's location – lying in a bowl of hills whose fast-flowing rivers could power waterwheels – made it perfect for large-scale metalware production when the Industrial Revolution got under way. The population grew tenfold during the nineteenth century. Sheffield became one of the world's great steelworking cities, and one of the shabbiest and slummiest. In the twentieth century, competition and lack of investment crippled the industry, but it still continues. Recently Sheffield, like Stoke-on-Trent, has been turning its derelict areas into parks, recreation places and arts venues.

*there and back*: Road: M1 (Leeds, Nottingham, London), A57 (Manchester).

Rail: Sheffield.

*go there for*: public parks, gardens and woodlands.

*festivals*: Sensoria Festival of Music and Film (April); Broomhill Festival (June); Sheffield Comedy Festival (October).

*mad fact*: Robin Hood of Sherwood Forest, most famous figure in Nottinghamshire folklore, was actually Sheffield born and bred.

*on the way to*: Dark Peak district; Sherwood Forest; Little John's Grave at Hathersage; Chatsworth.

*psychic smell*: leafy.


### Leeds

*population*: 770,000.

*football team*: Leeds United FC ('The Whites, The Peacocks') at Elland Road.

*bands*: Kaiser Chiefs, Gang of Four, The Mission.

*famous folk*: Alan Bennett (1934– ), actor/director; Chris Moyles (1974– ), disc jockey; Barbara Taylor Bradford (1933– ), novelist; Ellery Hanley (1961– ), rugby league player; Vic Reeves (1959– ), comedian.

*famous joke*: Long-suffering Leeds fan 1: 'What's the difference between a pyromaniac and us?'

Long-suffering Leeds fan 2: 'A pyromaniac doesn't chuck away all his matches.'

*matchbox history*: With Kirkstall Abbey's wide estates of sheep-friendly wolds (rolling hills) at hand, the medieval market town of Leeds became a centre for making and trading woollen cloth. Rivers tumbling off the nearby moors powered the looms of the Industrial Revolution, and the advent of steam power, canals and railways encouraged heavy engineering to develop in the hugely expanded city. Leeds became a byword for production, and for its classic 'dark satanic mills' and packed slum districts. Nowadays, post-industrial Leeds thrives on electronic and cyber-industries, and people choose to live along the once-filthy and pongy canals.

*there and back*: Road: M1 (Sheffield, London); M62 (Hull, Manchester).

Rail: Leeds.

*go there for*: fabulously overblown Victorian 'civic pride' architecture; mills and factories, grim or elaborate.

*festivals*: Leeds Beer, Cider and Perry Festival (March); Leeds Festival (August: rock music); International Film Festival (November).

*mad 'fact'*: Leeds is haunted by spectral hounds called Gabble Ratchets, and by demonic black dogs known as Barguests.

*on the way to*: Kirkstall Abbey; Harewood House; Ilkley Moor; York.

*psychic smell*: dark.

## North-West

*Liverpool*

*population*: 430,000.

*football teams*: Liverpool FC ('The
    Reds') at Anfield; Everton FC
    ('The Toffees') at Goodison
    Park.

*bands*: The Searchers, The Coral, Echo and the Bunnymen,
    The Teardrop Explodes, The Merseybeats, Atomic
    Kitten, The La's, The Zutons, The Sonic Hearts,
    Amsterdam, Frankie Goes To Hollywood . . . oh, yeah,
    and The Beatles.

*famous folk*: John Lennon (1940–1980), Paul McCartney
    (1942– ), George Harrison (1943–2001), Ringo Starr
    (1940– ), musicians; Robert Runcie (1921–2002), Arch-
    bishop of Canterbury; Wayne Rooney (1985– ), footballer;
    Glenda Jackson (1936– ), actress and politician; Ken
    Dodd (1927– ), comedian.

*famous joke*: Monkey and Scouser fly to the moon. Before
    they start, Monkey opens his sealed orders, and finds a
    rocket pilot's instruction manual, 1,000 pages thick,
    marked 'Read and Use'. Scouser opens his sealed orders.
    They say: 'Feed the Monkey.'

*matchbox history*: Like Bristol, Liverpool faces west towards
    the Americas; unlike the West Country city, Liverpool
    has the deep-water Mersey estuary on its doorstep. Trade
    with the seventeenth- and eighteenth-century colonies
    across the Atlantic; nineteenth-century mass emigration
    and immigration, especially from Ireland and on to
    America; twentieth-century transatlantic liners and huge
    cargo ships: Liverpool accommodated and prospered on

them all. Cosmopolitan, multi-ethnic, full of characterful music and people, even late twentieth-century seaport decline and urban decay haven't kept this great Northern city down.

*there and back*: Road: M62 (Manchester, Leeds), M58 to M6 (Birmingham and the north-west).

Rail: Liverpool Lime Street.

*go there for*: fab music, past and present; great imperial architecture; Merseybeat locations; Scouse wit and backchat.

*festivals*: Liverpool Beer Festival (February); Liverpool Performing Arts Festival (March); Liverpool Comedy Festival (May); Liverpool Sound City (May); Liverpool International Theatre Festival (May); Summer Pops (June/July); Matthew Street Festival (August), and many more.

*mad 'fact'*: An Internet scam has given rise to the widespread belief that a Liverpool bylaw outlaws all topless saleswomen, except those working in tropical fish shops. Well, of course . . .

*on the way to*: The Wirral peninsula; Antony Gormley's 'Another Place' art installation of 100 iron men on Crosby Beach.

*psychic smell*: salty.


## Manchester

*population*: 465,000.

*football teams*: Manchester United FC ('The Red Devils') at Old Trafford; Manchester City ('The Citizens, The Blues') at City of Manchester Stadium.

*bands*: Joy Division, New Order, The Happy Mondays, The

Smiths, The Fall, The Hollies, 10cc, The Stone Roses, The Buzzcocks, The Chemical Brothers, Elbow, Swing Out Sister, Take That.

*famous folk*: Mark E. Smith (1957– ), bandleader; Anthony Burgess (1917–93), novelist; Emmeline Pankhurst (1858–1928), women's emancipation activist; Noel Gallagher (1957– ), musician; L.S. Lowry (1887–1976), artist; Don Whillans (1933–85), mountaineer; Dodie Smith (1896–1990), playwright (*I Capture The Castle*) and novelist (*The Hundred and One Dalmatians*).

*famous joke*: Mancunian footballer in Chinese restaurant: 'Waiter, them noodles tastes a bit crunchy!'

Waiter: 'Them's the chopsticks, sir.'

*matchbox history*: How could Manchester fail? Close to the successive power sources of moorland rivers and coalfields, 'blessed' with the damp climate that cotton-spinning demands (otherwise the threads break), with a big work population and an outlet to the world in the nearby port of Liverpool, Lancashire's great industrial giant became the textile capital of the world. There was muck and brass a-plenty, till cheap foreign imports killed the trade. The grandiose red-brick and sandstone city that the Victorian merchants created went on the slide in the late twentieth century, but has experienced a style resurrection. Cool music and nightlife, great museums, an ever-popular university, shopping and entertainment complexes like Salford Quays, and the worldwide profile of . . . one of its football teams have combined to give the handsome city a tremendous shot in the arm.

*there and back*: Road: M60 (circular), M62 (Liverpool and Leeds), M66 (Bury), M61 (Preston and M6), M56 (North Wales, and M6 to Birmingham).

Rail: Manchester Piccadilly (London), Manchester Victoria (stations to north and east).

*go there for*: shopping, entertainment, clubs; football and cricket; a ride on a tram; musical history; 'Madchester' sites.

*festivals*: Manchester Irish Festival (March); Dot To Dot Festival (May); Manchester Gay Pride Festival (August); Manchester Comedy Festival (October).

*mad fact*: Greater Manchester produces 1.5 million tonnes of rubbish every year. If you built a skip big enough to hold it, you could throw in the Great Sphinx, too!

*on the way to*: Leeds, Liverpool and Sheffield; the Lancashire moors and football towns; the Dark Peak moors.

*psychic smell*: baggy.


## North-East

*Newcastle-upon-Tyne*
*population*: 275,000.
*football team*: Newcastle United FC ('The Magpies, The Toon, The Geordies') at St James's Park.

*bands*: The Animals, Sting, Pet Shop Boys, Dire Straits, Lindisfarne.

*famous folk*: Sid the Sexist (1979– ), *Viz* cartoon character; Ove Arup (1895–1988), civil engineer; Alan Shearer (1970– ), footballer; Catherine Cookson (1906–1998), novelist; Ant and Dec (1975– ), TV presenters; Eric Idle (1943– ), comedian; Dave Richardson (1948– ), traditional musician; Flora Robson (1902–84), actress.

*famous joke*: Q: What's the difference between a kanageroo and a kanageroot?

A: One's an incorrectly orthographised antipodean marsupial, and the other's a Geordie stuck in a lift.

*matchbox history*: The Romans sited the eastern end of Hadrian's Wall near the mouth of the River Tyne, and Newcastle grew from there. Timber and iron ore were close by for ship-building on the deep-water river; coal, too, from hundreds of local pits, for exporting by sea or use in industrial processes. No wonder the city grew, developing a maze of yards, foundries, bottle-works and twisting riverside alleyways. Now the yards are still, the pits closed, and Newcastle has found a new role as a city of the arts (galleries, dance, music) and as night-time Party Central for the whole North-East.

*there and back*: Road: A1M (London and Edinburgh), A69 (Carlisle).

Rail: Newcastle Central.

*go there for*: nights out; fanatical football; learning (well-regarded colleges and university); Newcastle Brown Ale ('Lunatic Broth'); great collection of trans-Tyne bridges; ripe local accent and dialect; Great North Run.

*festivals*: Evolution music festival (February); ScienceFest (March); Eat! Newcastle Gateshead food festival (June); Great North Run half-marathon (September).

*mad fact*: 150 million bottles of Newcastle Brown Ale are produced each year. None of the beer is brewed in Newcastle (it's done in Tadcaster, Yorkshire), and two-thirds of it is drunk abroad.

*on the way to*: Hadrian's Wall; Scottish Border; Northumberland National Park and beaches; Beamish Open Air Museum.

*psychic smell*: party.

## Durham

*population*: 40,000.

*football team*: Durham City AFC ('The Citizens') at the Arnott Stadium.

*bands*: The 4 Kinnells.

*famous folk*: Robert Surtees (1803–64), humorous novelist; Trevor Horn (1949– ), record producer; Alan Price (1942– ), musician; Elizabeth Barrett Browning (1806–61), poet.

*famous joke*: Geordie comedian 1: 'How do you confuse a Durham miner?'

Geordie comedian 2: 'Show him two shovels and tell him to take his pick.'

*matchbox history*: County Durham, famously, is rich in coal, and the city sits atop a layer-cake of sandstone, shale and thin coal seams. The River Wear has carved a great bend in this soft rock, isolating a high, narrow peninsula, with its own well-water supply, surrounded by the river on three sides. Can you imagine a place better situated for defence in the lawless Middle Ages? That's why the great castle and cathedral were built there. Sir Walter Scott summed it up perfectly: 'Half church of God, half castle 'gainst the Scot.' The medieval prince bishops of Durham, rich and powerful rulers, had their own army, mint and law courts. Nowadays Durham University maintains the reputation and prestige of this remarkable coalfield city.

*there and back*: Road: A1M (London and Newcastle-upon-Tyne). Rail: Durham.

*go there for*: superb Norman cathedral; views of the peninsula; rowing on the Wear; a great university education; a stottie cake from Greggs the bakers.

*festivals*: Durham Miners' Gala (July); Durham Brass Festival (July); Durham Book Festival (October).

*mad 'fact'*: William the Conqueror was so scared by what he saw when he opened the tomb of St Cuthbert that he leapt on his horse and didn't stop galloping until he had crossed the River Tees twenty miles away.

*on the way to*: the regenerated Durham coast; Weardale moors; Beamish Open Air Museum.

*psychic smell*: cobbled.

# SCOTLAND

## Borders

*Stranraer*

*population*: 11,000.

*football team*: Stranraer FC ('The Blues') at Stair Park.

*bands*: Darkwater.

*famous folk*: James Caird (1816–92), agricultural reformer; John Ross (1777–1856), polar explorer; Colin Calderwood (1965– ), footballer.

*famous joke*: Raith Rovers fan 1: 'How many Stranraer fans does it take to change a light bulb?'

Raith Rovers fan 2: 'Trick question – they don't exist.'

*matchbox history*: Facing sheltered water in the crook of Loch Ryan, Stranraer grew as a marketing hub and coasting harbour for the wide, remote south-west of Galloway. Trouble and opportunities in Ireland – only 30 miles (48 km) away across the North Channel – were the catalysts for Stranraer's boom as a transport centre: a military road got there around 1600 to take troops across the water, a proper harbour was built and expanded in the eighteenth century, and the railway arrived in the 1860s.

*there and back*: Road: A75 (Carlisle and M6).

Rail: Stranraer.

Sea: Stena Line (Stranraer), P&O (Cairnryan).

*go there for*: ferries to Ireland; lowly but passionate Scottish League football.

*festivals*: Galloway Music Festival (March); Portpatrick Folk Festival (September).

*mad fact*: Stranraer's name in Gaelic is *an t-Sròn Reamhar* – 'Fat-Nose'.

*on the way to*: Ireland; the Rhinns of Galloway peninsula.

*psychic smell*: backwoods.

## Lowlands

*Edinburgh*

*population*: 470,000.

*football teams*: Heart of Midlothian FC ('Hearts, Jam Tarts, Jambos') at Tynecastle Stadium;

Hibernian FC ('Hibs, Hibees, The Cabbage') at Easter Road, Leith.

*bands*: The Proclaimers, Bay City Rollers.

*famous folk*: Alexander Graham Bell (1847–1922), inventor of the telephone; Arthur Conan Doyle (1859–1930), creator of Sherlock Holmes; Chris Hoy (1976– ), Olympic gold medal cyclist; Muriel Spark (1918–2006), novelist; Sean Connery (1930– ), actor and 007; Harry Lauder

(1870–1950), singer; Walter Scott (1771–1832), novelist and balladeer; Deacon Brodie (1741–88), gangleader and gallows-builder; Marie Stopes (1880–1958), birth control pioneer; Norman MacCaig (1910–96), poet; Hannah Gordon (1941– ), actress.

*famous joke*: Edinburgh lady to Glaswegian woman: 'Here, my dear, we think breeding is everything.'

Glaswegian woman to Edinburgh lady: 'Aye, well, ye should try an' get a few outside interests too.'

*matchbox history*: Uncounted generations lived, prospered, fought and died on the volcanic Castle Rock before the medieval Edinburgh Castle was built there. With lookout crags at its back and a great sea firth at its feet, commanding the neck of land connecting the Borders and England with the rest of Scotland, Edinburgh was always a hugely important city. In 1492 King James IV moved the Scottish court from Stirling to Holyrood, just downhill from the castle, and Edinburgh became the national capital. With its handsome Georgian New Town, and its newfound pride in national devolution, Edinburgh's story is one of ongoing success. It's a really handsome city, with its superb Georgian New Town lying parallel to 'Auld Reekie', the old part of the city that descends the Royal Mile from the castle to medieval Holyrood Palace and the ultra-modern Holyrood Building, home of the devolved Scottish Parliament.

*there and back*: Road: A720 (circular); A1, A68 (Newcastle-upon-Tyne); M8 (Glasgow); A90 (Forth Road Bridge, M90 to Perth); M9 (Stirling and A9 to Perth).

Rail: Edinburgh Waverley (East Coast, London, and Edinburgh Haymarket station); Edinburgh Haymarket (Glasgow, and north and west Scotland).

*go there for*: architecture that's military (Edinburgh Castle), late medieval (Old Town, Holyrood Palace), Georgian (New Town) and modern (Scottish Parliament's Holyrood Building); festivals; great pubs of Rose Street; nightlife; university.

*festivals*: Edinburgh International Film Festival (June); Edinburgh Festival and Edinburgh Fringe Festival (August/September); Edinburgh International Book Festival (August); Mela Festival (music and culture – August), and many more – plus Hogmanay Celebrations (31 December/1 January annually).

*mad fact*: Follicly challenged seventeenth-century Edinburgh gentlemen resorted to plastering their shiny scalps with ashes of burned dove poo. It didn't work.

*on the way to*: Leith, trendy dockland quarter; Pentland Hills for walking; East Lothian coast; Forth Bridges; Glasgow.

*psychic smell*: Gothic.

### Glasgow

*population*: 580,000.

*football teams*: Glasgow Celtic ('The Bhoys, The Hoops, The Celts') at Celtic Park; Glasgow Rangers ('The Gers, Teddy Bears, Light Blues') at Ibrox Stadium; Queen's Park ('The Spiders, The Glorious Hoops') at Hampden Park;  Partick Thistle ('The Jags, The Harry Wraggs, Maryhill Magyars, Maryhill Margaritas') at Firhill Stadium.

*bands*: Franz Ferdinand, Deacon Blue, Wet Wet Wet, Belle & Sebastian, Orange Juice, Average White Band, Del Amitri, The Jesus and Mary Chain, The Fratellis, Primal Scream, Sensational Alex Harvey Band . . .

*famous folk*: Lulu (1948– ), singer; Alex Ferguson (1941– ), football club manager; John Logie Baird (1888–1946), inventor of television; Mark Knopfler (1949– ), guitarist (his name means 'Little-buttons'); Carol Ann Duffy (1955– ), Poet Laureate; James Watt (1736–1819), steam power pioneer; Billy Connolly (1942– ), comedian; Charles Rennie Mackintosh (1868–1928), architect.

*famous joke*: Edinburgh lady, watching Glasgow woman settle in a chair: 'Comfy?'

Glasgow woman: 'Govan.'

*matchbox history*: Like London, Glasgow grew up round a river crossing, the first ford coming inland up the Clyde. From an early medieval religious centre, Glasgow became a city that swelled like a frog in a fable, thanks to its west-facing situation. Rum, sugar, cotton and tobacco came from the West Indies, shipyards proliferated along the deep-water shores of Clydeside, and Glasgow grew prosperous enough to develop warehouses and offices as elaborate as palaces – as well as such slums as the Gorbals and Saltmarket, as desperate as any in Britain. The magnificent architecture is still there; the slums have been cleared, though outer Glasgow still contains pockets of genuine poverty.

*there and back*: Road: M8 (Edinburgh and Firth of Clyde); M77, A77 (Ayrshire coast); M74 (Carlisle); M80, A80 (Stirling).

Rail: Glasgow Central (Borders, London); Glasgow Queen Street (Edinburgh, north of Scotland).

*go there for*: strong humour and strong accents; splendid warehouse and office architecture; nightlife; music; festivals; art nouveau architecture of Charles Rennie Mackintosh.

*festivals*: Celtic Connections music festival (January); International Comedy Festival (March); Aye Right! Book Festival (March); Art Fair (April); Glasgow River Festival (July); Piping Live! International Piping Festival, and World Pipe Band Championships (August); Glasgow Fireworks (5 November); Glasgow Hogmanay (31 December/1 January annually).

*mad fact*: The first McDonald's 'restaurant' in Glasgow was boycotted . . . because it didn't sell Irn-Bru.

*on the way to*: Loch Lomond and the Trossachs; the west coast and Isles; Stirling and Perth; Edinburgh.

*psychic smell*: in ya face, pal, get a problem wi' that?

## Highlands

*Inverness*

*population*: 70,000.

*football team*: Inverness Caledonian Thistle FC ('The Caley Thistle, The Caley Jags') at Caledonian Stadium.

*bands*: 11 Mile Drive.

*famous folk*: Charles Kennedy (1959– ), politician; James Swinburne (1858–1958), plastics pioneer; Elspeth Grey (1929– ), actress.

*famous joke*: Inverness lad: 'Why can't I swim in Loch Ness, Mam?'

Mam: 'Because there's a monster in there.'

Lad: 'Aye, but my daddy's swimming there.'

Mam: 'Aye, but he's insured.'

*matchbox history*: An ancient stronghold of the Pictish people, Inverness was always going to be an important place, thanks to its sheltered position on the inner Moray Firth, its command of the northern mouth of the Great Glen, and its situation between Lowlands and Highlands. The city was often attacked, the castle taken and the surrounding ground contested between English and Scots, Scottish loyalists and insurgents, and local clans. Great forts were built on the Moray Firth to protect the city. The Caledonian Canal and the railway brought nineteenth-century prosperity to Inverness; the cutting of the canal made an island of the west side of the city.

*there and back*: Road: A82 (Glasgow), A9 (Edinburgh, north-east of Scotland), A835 (north-west of Scotland).
Rail: Inverness.

*go there for*: river and canal walks; street bagpipe music; Loch Ness cruises; Battlefield of Culloden.

*festivals*: Marymass Fair (June); Inverness Summer Festival (June–September); Highland Food & Drink Festival (September/October).

*mad 'fact'*: Pizza Express on Eastgate Mall is haunted by a ghost who pops the lightbulbs, blows the fuses, and opens and closes the doors.

*on the way to*: Loch Ness; north-west Highlands; Culloden Battlefield.

*psychic smell*: tartan.

# WALES

## Cardiff

*population*: 325,000.

*football team*: Cardiff City FC ('The Bluebirds') at Cardiff City Stadium.

*bands*: Catatonia, Super Furry Animals (from the Valleys: Lostprophets, Tom Jones, Manic Street Preachers, Stereophonics, John Cale).

*famous folk*: Charlotte Church (1986– ), singer; Colin Jackson (1967– ), Olympic hurdler; Gillian Clarke (1937– ), poet; Ivor Novello (1893–1951), songwriter and actor; Shirley Bassey (1937– ), singer; Howard Spring (1889–1965), novelist; Tanni Grey-Thompson (1969– ), Paralympic gold medal athlete; R.S. Thomas (1913–2000), poet and priest.

*famous joke*: Q: What does a Welshman call an Englishman who's opening a bottle of champagne after a win at the Millennium Stadium?

A: Waiter!

*matchbox history*: Cardiff on its tidal bay was a quiet coastal town of 6,000 inhabitants until the 1840s, when the biggest landowner in South Wales, the Second Marquess of Bute, developed the docks to handle the coal and iron pouring out of the Valleys just to the north. He had what amounted to a monopoly of import and export tolls. Within half a century Cardiff's population had swollen thirty-fold, to over 170,000. 'Bute Town', the dock area, made the city prosperous, and the Bute family rich beyond imagining. Now iron and coal shipping are gone, and the dock area is revamped as the Cardiff Bay development – with shops, bars, flats, nightlife, offices, and the Senedd, the devolved Welsh Assembly building.

*there and back*: Road: M4 (London and Swansea).

Rail: Cardiff Central (London, West and North Wales, main-line stations in England); Cardiff Queen Street (Welsh Valleys).

*go there for*: Cardiff Castle, the Bute family's eccentric, overblown residence; Cardiff Bay; strolling and eating; Llandaff Cathedral; summer festivals.

*festivals*: Mas Carnival (July); Welsh Proms (July); WOW on the Waterfront, street theatre festival (July); Cardiff International Food and Drink Festival (July); Admiral Cardiff Big Weekend, free outdoor music festival (July/August); Cardiff Harbour Festival (August).

*mad fact*: Llandough Hospital boasts the longest hospital corridor in Europe; it's a mile from end to end.

*on the way to*: the Welsh Valleys; West Wales; Bristol Channel; England.

*psychic smell*: resurgent.


## Swansea

*population*: 230,000.

*football team*: Swansea City FC ('The  Swans, The Jacks') at Liberty Stadium.

*bands*: Bonnie Tyler, Man, Badfinger.

*famous folk*: Catherine Zeta Jones (1969– ), actress; Dylan Thomas (1914–53), poet and playwright; Harry Secombe (1921–2001), singer and comedian.

*famous joke*: Your sister is your mother,
Your brother is your father,
You all 'love' each other,
The [insert rival team] family.
Come on, City, come on, City,
Come on, City, come on, City,
Come on, City, come on, City,
Come on, City, come on, City.
And it's Swansea City,
Swansea City FC,
We're the greatest team
The world has ever seen.

*matchbox history*: 'Svein's Island' at the mouth of the River Tawe was a Viking raiding base, then a Viking settlement. Come the Industrial Revolution, the town was perfectly placed, with coal at its back, deep-water anchorage at its feet, the Bristol Channel on its doorstep. Zinc, silver, brass, iron were all smelted in Swansea, but above all copper – by the nineteenth century, nine-tenths of all the world's copper was being smelted here in 'Copperopolis'. As Swansea smelted, so it polluted itself with the by-products of the process, arsenic and sulphur. A Lower Swansea Valley smelter wouldn't expect to see his twenty-fifth birthday. The river was horribly fouled, the ground poisoned. But a huge late-twentieth-century clean-up has seen the area largely restored.

*there and back*: Road: M4 (Cardiff).
Rail: Swansea.

*go there for*: Dylan Thomas Museum; revitalised waterfront.

*festivals*: Swansea Bay Summer Festival (May–September); Swansea Bay Film Festival (May); Gower Walking Festival, Gower Folk Festival (June); Swansea Angling Festival (July).

*mad fact*: Swansea is the wettest city in Britain; it has 53 in (135 cm) of rain a year – enough to drown a fair-sized man standing up.

*on the way to*: Pembroke National Park; Gower Peninsula; Cardiff.

*psychic smell*: chippy.

# NORTHERN IRELAND

### Belfast

*population*: 270,000.

*football teams*: Cliftonville Football and Athletic Club ('The Reds') at Solitude; Crusaders FC ('The Hatchetmen, The Crue') at  Seaview; Glentoran FC ('The Glens, The Cocks 'n' Hens') at The Oval; Linfield Football and Athletic Club ('The Blues') at Windsor Park.

*bands*: Them, Gary Moore, Van Morrison, Snow Patrol, Stiff Little Fingers.

*famous folk*: C.S. Lewis (1898–1963), writer; Mary McAleese (1951– ), politician and president of the Irish Republic; Bobby Sands (1954–81), politician and hunger striker; George Best (1946–2005), footballer; John Stewart Bell (1928–90), nuclear physicist; Mairead Corrigan (1944– ), Nobel Prize-winning peace campaigner; Van Morrison (1945– ), musician.

*famous joke*: Dublin man goes skydiving in Belfast at the weekend. Parachute fails to open, he lands in hospital with broken everything. Nurse: 'You should have asked around, anyone could've told you – nothing opens in Belfast on a Sunday.'

*matchbox history*: As with so many coastal towns, Belfast's fortune derived from its geographical situation – at the narrow end of Belfast Lough, with deep water for docks and shipyards, and its main market, mainland Britain, straight ahead across the Irish Sea. Ship-building (the *Titanic*, among thousands of vessels), linen-making (Belfast's nickname was 'Linenopolis'), tobacco, rope-making, machinery – Belfast prospered. Then the Troubles between Loyalists and Nationalists kicked off in the late 1960s, and the ensuing thirty years of strife saw investment diminish, firms pull out and prosperity dwindle. Since the Good Friday Agreement of 1998, there's been huge investment, massive grant aid, and a slow, jerky return to normality.

*there and back*: Road: M1 (Dublin); A2 (Ards Peninsula, Antrim Coast; M2, A6 (Derry).

Rail: Belfast Central (Dublin, Derry); Belfast Great Victoria Street (local services).

*go there for*: river trips; pub crawls; Botanic Gardens; black taxi tours of the 'political' (i.e. sectarian) murals; festivals; intense black humour; giant cranes Samson and Goliath.

*festivals*: Féile an Earraigh music festival (February); Belfast Film Festival (April); Festival of Fools (April/May); Belfast City Carnival (June); Belfast Pride Festival (July); Belfast City Blues Festival (August); Belfast Children's Arts Festival (October).

*mad fact*: 'Albert's got the time, *and* the inclination' – Belfast's comment on the city's Albert Memorial Clock tower, which leans 4 ft (1.2 m) out of the perpendicular.

*on the way to*: Antrim Coast and Giant's Causeway; Mourne Mountains; Strangford Lough.

*psychic smell*: dark and tasty.

## Derry

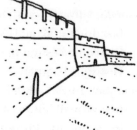

*population*: 90,000.

*football team*: Derry FC ('The Candystripes') at Brandywell Stadium.

*bands*: The Undertones, That Petrol Emotion.

*famous folk*: Joyce Cary (1888–1957), novelist; John Hume (1937– ), politician and Nobel Peace Prize winner.

*famous joke*: Outsider: 'You won't believe this, but I've heard that in Derry, two positives make a negative.'

Derryman: 'Aye right I won't.'

*matchbox history*: 'Stroke City' (because would-be-sensitive outsiders often refer to it as 'Derry-stroke-Londonderry') lies where the River Foyle widens into sheltered Lough Foyle. An ancient Celtic settlement, it was founded as the city of Londonderry in the early seventeenth century by trades guilds from London who'd been brought across to help establish a Protestant, Anglo-centric Ulster. They built the famous city walls that still encircle the town centre, and they brought trade and success. During the twentieth century, much of the Catholic population was reduced to living in slum conditions in communities such as the Bogside, outside the city walls. Their civil rights agitation was one of the triggers for the Troubles. Nowadays 'Stroke City' is a notably idiosyncratic, forward-looking place.

*there and back*: Road: A2 (Antrim Coast); A6 (Belfast); A5 (Omagh).

Rail: Londonderry/Derry.

*go there for*: walks round the city walls; guided walks through the Bogside; music; sardonic optimism.

*festivals*: City of Derry Drama Festival (March); City of Derry Jazz and Big Band Festival (April/May); Foyle Days (June); Celtronic Festival of electronic music (June); Maiden City Festival (August).

*mad fact*: In St Columb's Cathedral you'll find 'the world's first airmail letter' . . . a cannonball with a hole in it, fired in among the Protestant defenders during the epic Siege of Derry of 1688–9, with the Catholic besiegers' surrender demands written on a piece of paper and stuck into the cavity.

*on the way to*: Lough Foyle; Co. Donegal; Downhill Palace.

*psychic smell*: breezy.

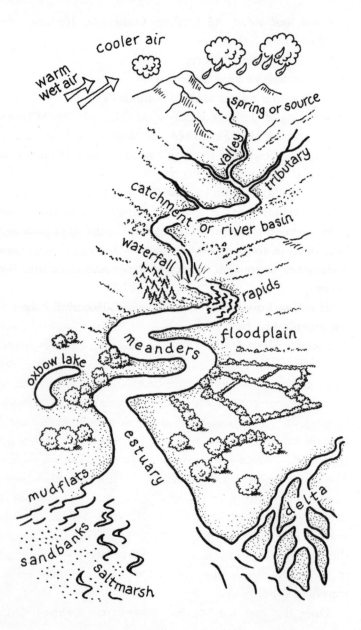

cooler air

warm wet air

spring or source

valley

tributary

catchment or river basin

waterfall

rapids

floodplain

meanders

oxbow lake

estuary

mudflats

delta

sandbanks

saltmarsh

# 7

## THE WATERY BITS

Compared with much of the rest of the world, we live in a Garden of Eden that's green and pleasant because it's very well watered by rivers. Here's some basic stuff about rivers:

The Gulf Stream currents warm the seas off the west of our archipelago. Most of Britain's wind comes from the south and west, over the Atlantic Ocean and across those warm seas towards the British Isles. It arrives laden with moisture it has picked up on its journey over the sea. When it meets the first rise of ground, the warm wet wind is forced upwards to meet the cooler air higher up. This condenses the moisture on the wind into clouds – that's why you so often get clouds over hills and mountains, as every hill walker knows. The cloud's temperature is lower than the open sky around it, so the moisture goes on condensing until it forms larger, heavier drops which saturate the cloud. Then the force of gravity causes them to fall earthwards as rain. Lots of it, in the case of our wet but fortunate **archipelago** – especially in the western hills, the first rise of ground that the east-travelling wind meets.

Once the rain has hit the ground, its seaward journey

begins. First of all it trickles over whatever bare rock may be there, before seeping into the first soil it encounters. From there it finds its way down through the soil and roots, between grains of grit, through cracks in the rock, to emerge at a weak place – a chink in the rock or an area of softer soil – as a **spring**, the **source** or starting place of the river.

Tumbling down from its source, the infant river cuts out a channel or valley – a '**dale**' in the north, e.g. Borrowdale in the Lake District; in Scotland a '**glen**' (narrower and higher up the river, e.g. Glenfeshie, the glen of the River Feshie that falls into Strathspey in the Cairngorms) or '**strath**' (broader and lower down, e.g. Strathspey itself, the strath of the River Spey); a '**cwm**' in Wales, e.g. Cwm Rhondda, the Rhondda Valley. It's incredible to see just how wide and deep a valley can be cut by what seems an insignificant little river – but don't forget it's had many thousands of years to do it. And if free-flowing water is a delicate sculptor of landscapes, frozen water in the form of ice is a demolition man with a hammer in both fists, and steel boots on as well.

*

**Intermission for Ices**: Climate change is absolutely nothing new. For example, the world has experienced five Ice Ages that we know about, and we're actually in one at present. Within each Ice Age there have been freezing (glacial) and less cold (interglacial) periods, and we're in the latest inter-glacial now. The next glacial freeze-up is almost certainly out there in the future somewhere, although it could be delayed by the effects of global warming.

During the last glacial period (it ended about 10,000 years ago), enormous ice sheets, or glaciers, went inching across northern Britain, some of them well over 1 mile (1.6 km) thick. The weight of these vast conveyor belts of ice crushed, squeezed and squashed the landscape beneath them, gouging out U-shaped valleys. The valley that Pennine Way walkers look down into from High Cup above Appleby-in-Westmorland is a great example. The glaciers also picked up and carried along huge boulders, scoring deep scratches in the rock surfaces as they ironed them flat. When these ice behemoths eventually began to melt and retreat northwards as the climate warmed up, they dumped the boulders (known as **erratics**, because they've wandered from their place of origin) all over the landscape. You can still find them in many places, completely at odds with their surroundings – for example, a clutch of smooth, speckly granite boulders, originally from around Shap in the north Pennine hills, which a glacier transported right across the country before releasing them on the beach at Sewerby near Bridlington on the east Yorkshire coast – well over 100 miles (160 km) away.

And now back to the play . . .

*

The river travels downhill, collecting **tributaries** – lesser trickles and streams ('**beck**' or '**gill**' in the north, '**burn**' in Scotland, '**nant**' in Wales) – as it goes. These tributaries, and the main river itself, drain the water from a wide area known as a **catchment** or **river basin**. For example, the River Medway, which flows through Kent, drains about two-thirds of the entire county, with a catchment of almost 1,000 square miles (2,600 sq. km) – six Rutlands, on the celebrated CRCS (County of Rutland Comparison Scale). West-flowing rivers in the west of England and Wales tend to be faster, more turbulent

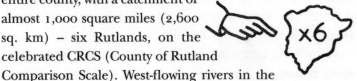

and cleaner than east-flowing ones, because they come from higher, hillier sources and water pours into them more directly off the bare mountain rocks. Here's where you'll most likely find **waterfalls**, such as the 240-ft (73-m) Pistyll Rhaeadr (Wales's tallest fall) on the upper Afon Disgynfa in the Berwyn Hills, and also **rapids** (water tumbling down 'staircases' in the riverbed – great for bold canoeists!) like those in the limestone bed of the River Ribble in North Yorkshire. East-flowing rivers, by contrast, often have their water thickened and polluted by heavy, clogging soil impregnated with agricultural chemicals picked up as they wind through the sticky, corn-growing claylands of the Midlands and East Anglia – Great Ouse, Welland and Nene, for example, which all flow to the Wash Estuary where Norfolk and Lincolnshire meet. That's why the water in the river mouths and the sea is clearer and more inviting to swimmers in Cornwall, on the west coast, than it is in, say, Essex on the east coast.

The more water and **silt** (material in solution) a river has to cope with, the more slowly it flows. The flatter the countryside, the more difficulty the river has in keeping up forward momentum – so it begins to snake around through its **floodplain** (flat country on either side, formed of silt spread during floods) in long bends known as **meanders**. The force of the river, pushing at the start of the bend, can sometimes cut a new, straight channel across the neck of the meander, isolating the old bend as an **oxbow lake**, a curved body of water left lying on its own with no connection to the river – as where the River Severn meanders through a flat floodplain at Caersws, not far from its source in mid-Wales.

As the river meets the sea it spreads out wide in the form of an **estuary** (a tidal channel, such as the Severn Estuary between the West Country and South Wales), or maybe a

**delta** (a branching web of channels, as on the Wash). Here's where the river tends to dump most of the soil and gravel it's been carrying, building up **sandbanks** or **mudflats**, or sometimes an apron of **saltmarsh** – rich silt, colonised and held together by tough, long-rooted plants that can stand salt, drought and regular drowning. These silty obstacles can choke off transport routes if they're not cut through or cleared away – as with former North Norfolk ports such as Cley-next-the-Sea and Blakeney, which now gaze across a mile of marshes at the distant sea that once brought them their livelihood.

\*

A river is always and everywhere a destroyer, breaking down its banks, crumbling its bed, washing away trees and rocks, overflowing and invading the surrounding land. It erodes and abrades. Its power is truly awe-inspiring. Given enough rainfall it can roll along boulders the size of a car, smash a house to bits or ruin a town (Lynmouth, Devon in 1952; Boscastle, Cornwall in 2004; Tewkesbury, Gloucestershire in 2007). Given enough time it will wear away the very mountains it comes from, dissolve them into particles too small to see, and hurry them off to drop them elsewhere.

These perfectly natural processes aren't 'bad'. A river builds, creates and enriches, too. Just like the sea, it may be eating away at its banks or its bed here, but it will be laying down a new sandbank or pebble island there. It cleans up worn-out or dead material, animal and vegetable, carrying it off to add it to the soup of life that we call the sea, or storing it along the riverbed as mineral-enriched silt until spring floods return this natural fertiliser to the floodplains around the river.

It's a cycle – moisture rising from the sea (and from lakes, reservoirs, streams, ponds, transpiring plants, peeing dogs and perspiring humans) to the hills on the wind, and descending from the hills to the sea as river water.

*

We have always used rivers – for transport, as a line of defence, for drinking, washing and cooking, for watering animals and crops, for fishing and fowling, for the fertile silt they spread during flooding to nourish grass and corn, for sport and play and general 'messing about on the river'. And we've probably always abused them, too, peeing and pooing in them, throwing our potato peelings and dead dogs and bust-up chairs into the water to be swept conveniently off elsewhere. With a small, widely scattered population this didn't matter too much, until the eighteenth century, when the Industrial Revolution added coal slag, iron ore, chemicals, the rubbish and human waste of crowded cities, and the run-off from tanning and smelting and ink-making and God knows what other industrial processes to the dumping ethos. And don't even mention the farm chemicals that came along last century. Actually, do mention them, because their nitrates have '**enriched**' the rivers and lakes that receive the run-off (accumulated water) from the farmland. Sounds a beneficial process, doesn't it? Who wouldn't want a richer river? But this kind of enrichment feeds the algae that turn rivers, lakes and ponds into thick soup, killing off the plants that should be in the river – and decimating the snails, shrimps, beetles, bugs and fish that feed on them, and on each other.

The Norfolk Broads, a string of flooded peat diggings smack in the middle of an intensively farmed area, went through this miserable process, from being clear and full of plant life at the turn of the twentieth century, to almost complete sterility as the millennium approached. And there are still many grossly polluted rivers in Britain. One of the worst, unsurprisingly, is the Severn, the longest river in Britain, with many towns and lots of farmland along its 220-mile (350-km) course, and an estuary lined with docks, chemical plants, industrial units, sewage outfalls and nuclear power stations. The River Trent is another bad 'un, flowing from its Staffordshire moorland source through the polluted Potteries region, the industrial clusters of Burton-on-Trent and Nottingham, and then a long stretch across the farmlands of Lincolnshire to its meeting with the River Ouse at the source of the River Humber, 170 miles (270 km) away. Sewage, chemical spills and farm run-off are the problem, as they are with the Dee (70 miles/112 km) on the England/Wales border, and the Bann (76 miles/122 km), which crosses Northern Ireland.

It's not all downhill. In the case of the Broads – given National Park status in 1989 – a big clean-up operation has been under way for several years. The River Lagan in Belfast, formerly a byword for disgusting smells of raw sewage at low tide, has been given a weir and better waste disposal, which between them have sorted out the problem. And the Thames, England's iconic river – so full of floating corpses and faeces in Victorian times that all the fish fled London, and Parliament had to abandon its riverside buildings because of the 'Great Stink' – is now largely cleaned up, thanks to better sewers and tighter controls.

Pollution and silting aren't the only troubles to visit our rivers. Flooding has always been a problem for people living near large, active rivers, which is why you won't find many old towns and villages on the most vulnerable floodplains –

they tend to be up on the slopes out of the way. But late in the twentieth century, an expanding population, government house-building targets, careless planning and developer greed led to building on floodplains, especially in the crowded south-east of England. Unprecedented cloudbursts and thunderstorms in 2007 unleashed devastating floods, which invaded towns as far apart as Hull, Sheffield, and Tewkesbury in Gloucestershire. An aerial photo of Tewkesbury Abbey, apparently floating in a lake of chocolate, gained national currency. With extreme weather events seemingly becoming more common, some authorities and developers have backtracked on floodplain building – for example, scrapping plans in 2010 for 7,500 houses on Kennet Meadows, a floodplain outside Reading in Berkshire. But others are still in the pipeline.

Other problems that dog the UK's rivers are over-extraction of water for an ever-increasing number of baths, showers, dishwashers and toilets up and down the land, and – sign of an ever-contracting global village – the spread of alien species. Red signal crayfish (brought here in the 1970s from America) are pushing out our native white-clawed crayfish. Mink, introduced for farmed fur in 1929, compete against the otter. New Zealand pigmy-weed and floating pennywort (North America), imported for ornamental ponds, have escaped into the river and pond system and are thriving at the expense of native plants as they compete for light and oxygen.

The Environment Agency is responsible for our rivers, and does a pretty good job in circumstances that often look like a stacking-up of insuperable odds. If the rivers of Britain were an index of the general environmental health of these islands, then the prognosis for the patient would have to be:

- used to be fit when young
- has recently been very sick
- is getting better . . .
- . . . but still has a long way to go.

Here are a dozen of the best . . . and one for luck.

## ENGLAND AND WALES

Rivers flowing east to the North Sea
tend to be muddy rivers, draining
big areas of low-lying farmland . . .

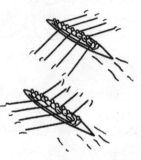

### River Thames
*Length*: 215 miles /346 km.
*Source*: Thames Head, near Kemble,
    Gloucestershire.
*Mouth*: east of London, between Essex and Kent.
*Course*: Thames Head (Gloucestershire), Cricklade (Wilt-
    shire), Lechlade (Gloucestershire), Oxford, Abingdon,
    Wallingford (Oxfordshire), Reading (Berkshire),
    Henley-on-Thames (Oxfordshire), Maidenhead, Windsor
    (Berkshire), London, Thames Estuary (Kent, Essex).

The Thames is the longest river in England, the
country's symbolic and ceremonial waterway, the river with
most of the royal connections. It springs from a modest
pile of stones in a Gloucestershire meadow, flows through
beautiful towns and villages of silver-gold Cotswold limestone, and threads classic Home Counties water meadows
and pastoral scenes (the settings for Jerome K. Jerome's

*Three Men in a Boat* and Kenneth Grahame's *The Wind in the Willows*) before passing right through the heart of London ... Windsor Castle, Eton, Hampton Court, Richmond, Kew Gardens, Varsity Boat Race territory, Westminster Abbey, Houses of Parliament, Big Ben, London Eye, South Bank, Tate Modern, Tower of London, Tower Bridge, The Gherkin, Canary Wharf, Millennium Dome, Thames Barrier ... a perfect roll-call of classic 'visitor's London'. And you can walk it all on the Thames Path National Trail. The Thames below London gets grittier, grimier and more mysterious – adventurous country for those who look and walk outside the box.

### Great Ouse

*Length*: 143 miles (230 km).
*Source*: near Sulgrave, Northamptonshire.
*Mouth*: The Wash, Norfolk.
*Course*: Brackley (Northamptonshire), Buckingham, Milton Keynes, Newport Pagnell (Buckinghamshire), Bedford (Bedfordshire), St Neots, Huntingdon, St Ives, Earith (Cambridgeshire), Downham Market, King's Lynn (Norfolk).

There are several UK rivers called Ouse (it just means 'water'), but only one Great Ouse – the fourth longest river in these islands, and the main navigable river in East Anglia. You can get a vessel up as far as Bedford, about 70 miles (112 km) inland of the Wash. 'Ouse' also means 'ooze', and that, too, defines the muddy and slow-flowing Great Ouse as it winds to and fro in largely flat country, absorbing clay and peat. The Great Ouse used to flood disastrously from both ends – sea surges from the Wash, freshwater floods from rainstorms in the hills surrounding its catchment (a mighty 1,320 square miles/3,420 sq. km – almost nine Rutlands on the CRCS). That was fixed when the Earl of Bedford got in Dutch water engineer Cornelius Vermuyden to bypass the winding Great Ouse and drain the sodden marshes of Fenland for agriculture. Vermuyden cut the Old Bedford River in 1630–36, then the New Bedford River in the 1650s, to take freshwater floods and marsh water to the sea. Water comes the other way, too – you can see the tide ripples in the New Bedford River far up its course. The two dead-straight artificial rivers, 20 miles (32 km) long and half a mile (800 m) apart, make a strange and unforgettable sight as they forge north-east across Fenland from Earith, where the diminished old Great Ouse channel wanders off east. It rejoins at Denver Sluice after 31 wriggling miles (50 km) in the back of beyond.

### River Humber

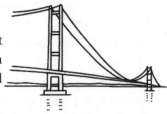

*Length*: 40 miles (65 km).
*Source*: Confluence of River Trent and Yorkshire Ouse, between Faxfleet (East Yorkshire) and Alkborough (Lincolnshire).
*Mouth*: Humber Estuary, between East Yorkshire and Lincolnshire.
*Course*: Humber Bridge, Kingston-upon-Hull (East Yorkshire), Grimsby (Lincolnshire).

Is the River Humber really a river, or is it an estuary? It's certainly tidal all the way seaward from its source, which is not a picturesque spring bubbling from a rock, but the **confluence** (merging) of two major rivers – the Yorkshire Ouse, which has travelled 100 miles (160 km) from its source (where it's known as the River Ure) above Wensleydale in North Yorkshire, and the River Trent, which has come 185 miles (300 km) from its source on the Staffordshire moors. Between them the huge catchments of the Trent and the Yorkshire Ouse drain about 6,000 square miles (15,000 sq. km) – forty Rutlands on the CRCS, or one-eighth of the whole of England. That is a hell of a lot of water pouring into the Humber. The broad tideway sweeps east under the Humber Bridge (nearly three miles from bank to bank), passes the

port of Hull and the oil refineries of Immingham, and reaches its mouth between the anteater snout of the Spurn Head peninsula to the north and the fish-processing town of Grimsby to the south. Here the Humber is 5 miles (8 km) wide, a leviathan of a river.

### River Tees

*Length*: 70 miles (113 km).
*Source*: Cross Fell, near Appleby-in-Westmorland, Cumbria.
*Mouth*: Teesmouth, between North Yorkshire and Tyne & Wear.
*Course*: Cross Fell (Cumbria), High Force Waterfall, Middleton-in-Teesdale, Barnard Castle, Darlington, Stockton-on-Tees, Middlesbrough (County Durham).

From its modest moorland birth to its polluted, smoky and workaday end, the journey of the Tees mirrors the progress of Industrial Britain. Upper Teesdale, the valley through which the young Tees flows, is a walker's and a naturalist's delight, a limestone dale into which intrudes the volcanic shelf known as the Whin Sill. This forms huge steps in the riverbed, over which the Tees tumbles – the tallest fall, High Force, is a stunning place when the river's in flood. Upper Teesdale also holds some very rare arctic-alpine flowers – royal-blue spring gentians, bright pink bird's-eye primroses, brilliantly coloured Teesdale violets – which have clung on here since shortly after the last glacial period 10,000 years ago. At the other end of its journey the mature River Tees slides past the ironworks, chemical plants and oil terminals of industrial Teesside, on out past Seal Sands and North Gare Sands (bizarrely, still breeding grounds for seals and seabirds) to meet the North Sea.

Rivers flowing south from the downs to the South Country coast tend to be chalky, clear and fast-flowing ...

### River Test

*Length*: 40 miles (64 km).
*Source*: near Ashe, Overton, Hampshire.
*Mouth*: Confluence with River Itchen in Southampton Water, Hampshire.
*Course*: Overton, Whitchurch, Stock-bridge, Romsey, Totton, Southampton (Hampshire).

The Test is a beautiful river, a classic trout stream. Its water springs from the chalk downs of north Hampshire, each spring and **rivulet** (small stream) already filtered clean by its passage through the lime-rich chalk by the time it joins the main flow of the Test. The river flows bright and shallow over a bed of gravel, shaded by willows and with many **back-waters** (waters diverted by opposing currents) and side streams – all conditions ideally suited for brown trout, which is why the Test along its upper waters is known as one of the great trout-fishing rivers of the world. Mills still draw water along the river – a paper mill producing paper for Bank of England banknotes, for example, and a silk mill in Whitchurch. Near its mouth, the Test threads industrial landscapes, passes the container port of Southampton and joins the River Itchen as it empties into Southampton Water.

Rivers flowing south and west to the Atlantic move *de haut en bas*, from the mountains to the sea . . .

### River Severn

*Length*: 220 miles (354 km).
*Source*: Plynlimon Mountain, Dyfed/Powys border.
*Mouth*: Severn Estuary at the Second Severn Bridge, between Gloucestershire and Gwent.
*Course*: Llanidloes, Newtown, Welshpool (Powys), Shrewsbury, Ironbridge, Bridgnorth (Shropshire), Bewdley, Stourport-on-Severn, Worcester (Worcestershire), Tewkesbury, Gloucester, Sharpness (Gloucestershire), Severn Bridges.

The mighty Severn is the longest river in Britain, and shares much of its course between England and Wales. Its unglamorous source is a muddy bog on the heights of Plynlimon Mountain in west Wales, and it thrashes around like a cornered snake as it makes its way east and north before finally settling on a purposeful south-ward flow. The Severn journeys from the sandstone of the mountains through a band of northern Welsh limestone, then the coal-bearing rocks of its gorge at Coalbrookdale, the cradle of the Industrial Revolution, where Abraham Darby first produced effective cast iron in 1709 (his beautiful bridge spans the gorge at Ironbridge, forming a perfect circle with its own reflection in the river). From there south it's a more pastoral journey through the low-lying plains of the Vale of Evesham, east of the Malvern Hills (very good for fruit-growing), and finally a spectacular series of tight bends through the silty clay and limestone of its Gloucestershire floodplain to empty into its long estuary. These bends are where you can get the best view of the Severn Bore, a tidal wave that travels regularly upriver, sometimes reaching ten feet in height.

### Afon Tywi

*Length*: 75 miles (121 km).

*Source*: 6 miles (10 km) north of Llyn Brianne, near Llanwyrtyd Wells, Dyfed/Powys border.

*Mouth*: Taf/Tywi/Gwendraeth Estuary, Carmarthen Bay, Dyfed.

*Course*: Tywi Forest (Dyfed/Powys), Llandovery, Llandeilo, Carmarthen (Dyfed).

The Afon ('River') Tywi is the longest river that flows wholly in Wales. It springs on the knobby hill of Crug Gynon ('mound of the brook', very appropriately) a mile or so west of Claerwen Reservoir in the back country of the Cambrian Mountains, and flows south through forestry and through Llyn Brianne before winding down in beautiful mountain scenery to its broader valley past

Llandovery and Llandeilo. It follows an edge of limestone west to Carmarthen, and then pours south through a sandstone gap into Carmarthen Bay by way of a great three-armed estuary. Here it meets and mingles with the Afon Gwendraeth coming from the east, and the Afon Taf flowing from the west past Laugharne and the home of Welsh national poet Dylan Thomas.

### Afon Rheidol

*Length*: 19 miles (31 km).
*Source*: Plynlimon Mountain, Dyfed/ Powys border.
*Mouth*: Aberystwyth (Dyfed).
*Course*: Ponterwyd, Devil's Bridge, Aberystwyth (Dyfed).

The Rheidol is one of those sparkling, energetic rivers that carves itself unforgettable scenery. It rises near the summit of Plynlimon Mountain, very near both Wye and Severn rivers, and flows south and then west to join the Afon Mynach and plunge down a succession of really spectacular waterfalls. At Devil's Bridge these are spanned by three bridges, one on top of another: a medieval bridge built by monks, an eighteenth-century bridge on top of that, and a twentieth-century iron bridge (the current roadway) spanning them both. Legend says the Devil built the lowest bridge for an old crone, on condition he could have the first living thing that crossed it (i.e. her). But she sent her little dog over first, and the evil one had to content himself with Fido. Below Devil's Bridge, what a contrast! The lively moorland river becomes grossly polluted with lead and zinc from abandoned mines up the slopes, and limps into Aberystwyth feeling very sorry for itself. You can travel this lower section on the Vale of Rheidol steam railway, looking down on the river from rock ledges far above.

### River Duddon

...in radiant progress...

*Length*: 15 miles (25 km).

*Source*: Wrynose Pass, Furness Fells, Cumbria.

*Mouth*: Duddon Estuary, Cumbria.

*Course*: Cockley Beck, Seath-waite, Ulpha, Duddon Bridge (Cumbria).

William Wordsworth was so moved by the simplicity and purity of the River Duddon's spring high in the Furness Fells that he composed a series of sonnets (published in 1820) in the river's honour. The Duddon's birth was innocent . . .

> Child of the clouds! remote from every taint
> Of sordid industry thy lot is cast;
> Thine are the honours of the lofty waste . . .

. . . as was its end in the wide Duddon Estuary 15 miles (25 km) to the south:

> . . . in radiant progress toward the Deep
> Where mightiest rivers into powerless sleep
> Sink, and forget their nature – now expands
> Majestic Duddon, over smooth flat sands
> Gliding in silence with unfettered sweep!

That was before the ironworks and iron mines opened at Millom on the estuary. By the time Wordsworth died in 1850, the Duddon Estuary ran red with poisonous haematite, glowed orange with furnaces day and night, rang with noise, choked in smoke and cinders. A century later, the industry was on the slide. Nowadays no iron is worked at Millom – little else, either. The Duddon once more glides unfettered in radiant progress toward the Deep.

## SCOTLAND

### North to the North Sea . . .

*River Spey*
*Length*: 107 miles (172 km).
*Source*: Loch Spey, Monadhliath
   Mountains, Highland.
*Mouth*: Spey Bay, Moray.
*Course*:   Garvamore,   Balgowan,
   Kingussie, Aviemore, Grantown-on-Spey (Highland), Charlestown of Aberlour, Rothes, Fochabers, Spey Bay (Moray).

Of Scotland's rivers only the Tay is longer than the Spey, and none is as spectacularly fast-flowing. Springing in the south-westerly tail of the Monadhliath Mountains (western neighbours of the Cairngorms), only a twist of the terrain prevents the Spey from flowing west into the Atlantic Ocean. Instead it sets off north-east, soon dropping into the long, straight strath (wide river valley) it has cut for itself between Monadhliath and Cairngorm. This is quartz country, with slate and granite – old, hard rocks, grooved with side glens that bring hundreds of burns of clean, peat-rich water down to the Spey. Whisky-makers need water just like this: hence the dozen or so famous malt whisky distilleries along Strath Spey – Glen-livet, Glen Grant, Cardhu, Glenfiddich, Knockando, to name a few. With its wide, fast water, the Spey is also famous as a salmon river, and you'll see anglers in waders, up to the waist in the river, as you journey north to the final tumbling estuary and a mighty shingle bank in Spey Bay.

## West to the Irish Sea . . .

### River Clyde

*Length*: 106 miles (172 km).
*Source*: Watermeetings near Daer Reservoir, Lowther Hills, South Lanarkshire, Central.

*Mouth*: Tail of the Bank, Greenock, Firth of Clyde.
*Course*: Crawford, Biggar, Lanark, Motherwell, Rutherglen, Glasgow, Clydebank (Central).

It's been a long time since the whole of working-class Glasgow would empty each July on Glasgow Fair and sail off 'doon th' watter' for a week. The 'watter' was the Firth of Clyde, the city's highway west to fun and frolics in boarding houses, pubs and music halls at Rothesay, Ayr, Greenock and Garelochhead. How many of those shipyard workers, rope-winders, dockers, hammer men, fleshers and bonnet-makers ever got themselves back in the Lowther Hills to where their familiar River Clyde springs in a mesh of burns you can hop across? The

Clyde's northward journey to its firth is through wild country at first; then it begins to squeeze in where steel-works, iron foundries, coal and chromium plants once ran together in the southern suburbs of Glasgow. A semi-derelict, dirty scene not so long ago, the Clyde's waterfront through the heart of the city is now a stretch of housing, bars, night spots and 'clean' businesses – only the start of the hugely ambitious Clyde Waterfront regeneration project. Watch that space . . .

## North to the Atlantic . . .

*Kervaig River*

*Length*: 5 miles (8 km).

*Source*: Loch na Gainmhich and Loch na Glaic Tarsuinn, Cape Wrath, Highland.

*Mouth*: Kervaig Bay, Highland.

*Course*: North by Kervaig Bridge.

One of the shortest, and one of the sweetest. The Kervaig River is nothing more than 5 miles (8 km) of ice-cold flow, charged with peat and heather particles, issuing from a moorland loch and snaking through bogland until it rushes down a cleft and out to sea over an apron of chattering pebbles and a hem of creamy, unsullied sands marked only by otter and gull prints. And where can you taste the joys of this paradise river? Ah – that's the catch. First you have to get yourself to Cape Wrath, the remotest part of north-west Scotland – in fact, the point at which the west coast of mainland Britain quits the Atlantic, wheels right and sets off for the North Sea. Then it's easy

– walk east along the lonely cliffs, and listen for the roar of the surf in Kervaig Bay. Fording the river barefoot to reach the sands is one of the great wild moments of walking in Britain.

# NORTHERN IRELAND

## Across the whole country ...

### Bann River

*Length*: 80 miles (129 km).

*Source*: Deer's Meadow near Spelga Dam, Mourne Mountains, Co. Down.

*Mouth*: The Barmouth between Castlerock and Portstewart, Co. Derry.

*Course*: *Upper Bann*: Spelga Dam, Hilltown, Ballyroney, Katesbridge, Banbridge (Co. Down), Portadown, Craigavon, Lough Neagh (Co. Armagh); *Lower Bann*: Lough Neagh, Toome, Lough Beg, Portglenone, Kilrea (Antrim/Derry border), Coleraine (Co. Derry).

The Bann is the longest river in Northern Ireland, a tranquil north-westerly flow for most of its length. Beginning, middle and end are well worth exploring. Any river that starts in a range of mountains as striking as the tall, shapely Mournes deserves homage at its birth, while the mouth of the Bann pushes out to sea between great lighthouse-tipped breakwaters flanked by long golden strands. What happens halfway along its course is rather extraordinary – the Bann enters the largest lake in Britain, the great inland sea of Lough Neagh (151 square miles/391 sq. km or one Rutland on the CRCS), on its southern shore in County Armagh, swims underwater for 20 miles (32 km), and emerges at the north end between counties Antrim and Derry like a gleaming green newt to continue its northward windings.

# 8

## BRITANNIA'S BULWARKS –
## COASTS AND ISLANDS

*Britannia needs no bulwarks,*
*No towers along the steep;*
*Her march is o'er the mountain wave,*
*Her home is on the deep.*

– Thomas Campbell in *Ye Mariners of England* (1800)

### COASTS

Maybe so, Thomas Campbell (sentimental Scottish poet, 1777–1844) – but, need them or not, Britannia has, mile for mile, just about the most varied and beautiful bulwarks in the world. Anyone who has watched BBC2's monumental series *Coast* unroll over recent years must have been struck by that. It's down to the remarkable geology of the British Isles, which contains a chronology of rocks that runs from some of the oldest to some of the youngest on earth – and to our possession of an in-and-out, up-and-down coastline that's 20,141 miles (32,414 km) long. (Pop stat: if you unrolled the UK's coastline and stretched it out flat, it would almost encircle the Earth.) It's an enormously long coastal strip, and a lot of it consists of cliffs, each of which allows several hundred million years of these islands' history to be exposed on one convenient, vertical, easy-to-read rock page. What luck for the fortunate Brits!

Another, related topic that recurs again and again in *Coast* is the extraordinary number of weird and wonderful towers along Britannia's steep; towers and castles, forts and pillboxes, lookouts, gun batteries, barracks, bunkers and blockhouses. That's all a function of being a little cluster of islands with an enormous – let's face it, a disproportionate – history. Britain's industrial might and her overseas trade may have dwindled, her Empire is no more, her colonies have shredded to the four winds, her voice is a whisper in the corridors of world power these days. But the five-hundred-year adventure has left us with something to treasure and marvel at, a wild coast that bristles with defensive structures, the stamp of a tiny archipelago that tyrants have wanted to subdue, a story of invasion scares and attacks, defiance and bloody-minded endurance.

But history is not written solely in blood and bullets. Many of those towers along the steep – lighthouses, fishermen's lookouts, signal stations, coastguard posts – bear witness to these islands' peacetime (though rarely peaceful) engagement with the sea. Walk or ride or canoe along Britannia's bulwarks, and at almost every place where the rocks are soft enough to allow a river to cut through them to the sea, or the sea to carve out a sheltered indentation, you'll find a port, a harbour, a seaside resort or a fishing village – or maybe just a row of coastguard cottages, on the alert for smugglers who have long passed into legend. If geography is the story of a place – and it certainly is, and how I wish Mr Matt had put it that way to me back in the classroom (perhaps he did, and I wasn't listening) – then the British coast tells the tale of an unending alliance and an unceasing enmity between man and sea.

More than any other single element, the sea shapes our island story. On the defensive side are great English naval ports such as Portsmouth, Chatham and Devonport, facing an often hostile Europe only a few miles away across the English Channel; Scotland's naval bases on the Clyde at Faslane (looking to the Atlantic) and at Rosyth on the Forth (facing into the North Sea). On the commercial front, great ports lie up the estuaries, receiving and dispatching goods and passengers: London on the Thames, Southampton on the Solent, Bristol and Cardiff on the Severn, Hull on the Humber, Liverpool on the Mersey, Newcastle on the Tyne, Glasgow on the Clyde, Edinburgh on the Forth. And dotted in between, hundreds of coastal trading towns and trawling harbours, fishing villages and tuppenny-ha'penny quarries and jetties and landing slips that bear witness to our ineluctable relationship with the sea.

## Seas

All you really need to know about the sea that surrounds the British Isles is that it's the **Atlantic Ocean** to the west, and the **North Sea** to the east. However, the various stretches of water that run between the different mainlands and islands have individual names, some well known, others surprisingly obscure. Going clockwise round Britain from the mouth of the River Thames:

• The short strip of Kentish coast that encompasses Deal, Dover, Folkestone and Dungeness is separated from the French coast by the narrow **Strait of Dover**, one of the world's busiest shipping lanes, with up to 500 ship

movements a day through the strait. Only 21 miles (33.6 km) wide between Dover and Cap Gris-Nez, the Germans developed a wartime gun that could fire right across it.

- From the Strait of Dover all the way west to Land's End and the Isles of Scilly, the diverging coasts of England and France are kept apart by the **English Channel**.

- From Land's End north-east to Morte Point in north Devon (where the coast turns east towards Somerset), you look out on the open **Atlantic.**

- From Morte Point up the English coast to the Severn Bridges near Bristol, and from there west along the Welsh coast out to the western tip of Pembrokeshire, it's the **Bristol Channel**.

- From western Pembrokeshire up the coast of west Wales as far as the Llŷn Peninsula, it's **Cardigan Bay**. The stretch of water that lies further out, between Wales and Ireland, is **St George's Channel** (why isn't it St David's Channel, I wonder?).

- From Anglesey north to the Rhinns of Galloway at Scotland's south-west tip, it's the **Irish Sea**.

- From Galloway as far as the Mull of Kintyre, the narrow **North Channel** (only 20 miles [32 km] wide) separates south-west Scotland from Northern Ireland.

- Northwards from the Isle of Islay (southern point of the Inner Hebrides archipelago), you look west into the **Atlantic**. But the long arc of the Outer Hebrides or Western Isles archipelago, 50 miles (80 km) further out, soon comes between the Inner Hebrides and the open ocean. The stretch of sea between the two archipelagos is called the **Little Minch** as far north as the top of the Inner Hebrides, then the **North Minch** from there north to the Butt of Lewis, the northernmost point of the Outer Hebrides.

- Across the top of Scotland it's the **Atlantic** – but the tide-ripped sea strait (narrow bit) between Dunnet Head and Duncansby Head (the north-east corner of the mainland) and the Northern Isles (the Orkney and Shetland archipelagos) is the **Pentland Firth**. If anywhere specific, it's here around the Northern Isles and Pentland Firth that the Atlantic Ocean meets and mingles with the North Sea.
- From Duncansby Head south-west to Inverness, and east again to Fraserburgh, the big triangle of sea enclosed by the coast is the **Moray Firth**, an inlet of the **North Sea**. And from there on, all the way down the east coast of Scotland and England as far south as the east Kent coast, it's the North Sea you look out on.

## Islands

Q: What is an island?

A: A piece of land entirely surrounded by water.

Ah, yes – but what do you mean by 'land'? Something the size of the Isle of Lewis/Harris in the Outer Hebrides (841 square miles = 2,178 sq. km = 5½ Rutlands = the largest island off mainland Britain), complete with people, towns, villages, agriculture and industry? Or a blob the size of Bishop Rock in the Isles of Scilly, with just enough room for a lighthouse? Well – both of these, and everything in between, count as part of the sprinkling of some 5,000 islands throughout the British archipelago.

The geology of the western and northern regions of the British Isles tends to be more volcanic than that in the east and south, so more knobs and bumps of rocky land have been pushed up in the west and north, to remain above the sea even when its level has risen from time to time. The rocks are generally hard and old, so they endure where softer material would be washed away. And the west coast also faces a much stronger, stormier and more energetic onslaught from the big, far-travelled Atlantic waves than does the east coast from the North Sea; so the sea's bites are deeper in the west and north, its capacity for cutting coasts into peninsulas, and peninsulas into islands, much greater.

For all these reasons you'll find the vast majority of our 5,000 islands in three archipelagos off the west and north coasts of mainland Scotland – Inner and Outer Hebrides to the west, Northern Isles to the . . . guess where?

Here's a clockwise journey before the mast round the islands of Britain, corralled into fifteen groups. Remember – port is left, starboard's right in sailor's lingo. OK? Let's start down and dirty, in the Thames estuary mud of south-east England . . .

## Thames Estuary

Sailing downriver from London, you pass two significant islands lying flat to port along the north or Essex shore: **Canvey Island** with its split personality (half empty grazing marshes, half tight-packed housing, lots of gas and fuel storage,

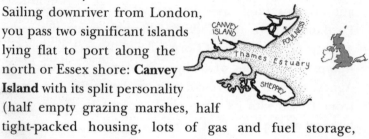

great R&B music courtesy of favourite sons Dr Feelgood), and **Foulness**, a 'closed' Ministry of Defence establishment. On your starboard hand lies **Sheppey**, with crumbling clay cliffs and flat marshes, and the marsh-and-creek islets of the Medway delta. So far, so moody, muddy and marshy . . .

*English Channel*

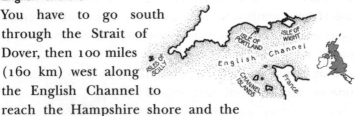

You have to go south through the Strait of Dover, then 100 miles (160 km) west along the English Channel to reach the Hampshire shore and the next sizeable island, the **Isle of Wight**, with rolling green fields, woods, villages and fine southern cliffs of chalk culminating to the west in the upstanding chalk blades called The Needles. On again to Dorset where the great lion-shaped block of freestone (hard-wearing but easily worked limestone) called the **Isle of Portland** sticks out into the sea. A plumb-line dropped south from Portland down the map for 60 miles (100 km) would knock against the **Channel Islands** – Jersey, Guernsey, Alderney, Sark and their off-islands – sheltering off the Cotentin peninsula of Normandy, neither French nor British, but their own autonomous (and tax-efficient) selves. West again from Portland, where the English Channel meets the wider Atlantic off the toe-tip of Cornwall, you'll reach the granite humps of the **Isles of Scilly** – St Mary's, St Martin's, St Agnes, Bryher and Tresco; remote, beautiful and (apart from Tresco with its subtropical garden) harshly scoured by wind and spray.

## Bristol Channel

Set course from the Scillies up
the north coasts of Cornwall
and Devon to reach **Lundy**,
a granite bar 11 miles
(18 km) offshore in the
throat of the Bristol

Channel. Once the 'Kingdom of Heaven' (it was owned by
the Heaven family), it's now a holiday place. Much further
up-channel you'll find round, humpbacked **Steep Holm** (a
nature reserve with black-backed gulls and rare wild peonies)
and appropriately low-lying **Flat Holm**, formerly a quaran-
tine island for sailors, these days a day-trip destination.

## Pembrokeshire

Westward Ho! down the
Bristol Channel, along
the coast of South
Wales to the western
county of Pembrokeshire,

possessor of a cluster of islands,
widely distributed. **Caldey Island** off Tenby, the most easterly, is home to pupping grey seals in autumn, and a community of gentle, chocolate-making monks all year round. Off
the tip of Pembrokeshire lie the bird reserves of **Skokholm**
and **Skomer** (puffins, razorbills and the world's largest colony
of Manx shearwaters – nearly 200,000 of them packed into
earth tunnels). Further west rises gannet-haunted **Grassholm**;
further north the nature reserve of **Ramsey Island** with its
colony of choughs, secluded on the far side of Ramsey Sound,
the wild and narrow tidal channel between island and
mainland.

*North Wales*

Square away north through Cardigan Bay to pass the tip of the Llŷn peninsula and **Bardsey**, shaped like a sleeping otter, final resting place of countless medieval pilgrims. Off the north-west shoulder of Wales rises the lumpy green island of **Anglesey**, the largest in Wales and England (nearly two Rutlands on the CRCS), with a mountainous north-westerly extension in Holy Island, and – of course – the village with the longest name in Britain, if not the world: Llanfairpwllgwyngyllgogerychwyrndrobwllllantysiliogogogoch, 'the church of St Mary in the hollow of the white hazel near a fierce whirlpool and St Tysilio near the red cave'. Unstick your jaw from that; then head east and then north up the Merseyside coast, passing the three diminutive sandstone islets of the **Hilbre Islands**, targets for intrepid walkers on the tidal sands of the Dee Estuary. There's a lot of dusky red sandstone from here on until you get well into Scotland.

*Great Sands*

North again, keeping well out to sea (huge shallow shoals over vast sandbanks ahoy!) until you round to in Morecambe Bay (but make sure it's at high tide – there's 10 miles/16 km of sandbanks exposed when the tide goes out) and head west again into the Great Sands of South Lakeland. Here off Barrow-in-Furness lies the long, slender island of **Walney**, shaped like the head of a double claw-hammer, sheltering in its southerly claw two tiny islands: **Roa** at the end of a long causeway, and **Piel** isolated out in the sands with its castle and Ship Inn. The landlord of the Ship is titled 'King of Piel' – monarch of a lonely tidal paradise. From Walney head west into the Irish Sea until the horizon brings up the head of Snaefell, crowning peak of the **Isle of Man**. Like the Channel Islands, Man keeps its own customs and laws, while retaining strong links with the British mainland through holiday-making, a tax-exile lifestyle, the famous and notoriously dangerous TT road races for motorbikes, and a landscape that's eerily like the mainland as it used to be – narrow lanes, flowery banks, neat fields and immaculate villages.

## Northern Ireland

From Man head north-west through the North Channel, with the Galloway coast of south-western Scotland to starboard and the volcanic drama of Northern Ireland's Antrim cliffs to port, until the L-shaped bar of **Rathlin Island** heaves into sight. Not that Northern Ireland's only offshore island seems L-shaped from the sea, but its characteristic 'iced cake' geology (pale 'cake' of chalky limestone, topped with dark basalt 'icing') – and its quarter-million screaming seabirds – tell you where you are, right enough.

## Firth of Clyde

Now your course is shaped due east, passing the cliffs of the Mull of Kintyre to round into the broad mouth of the Firth of Clyde. No need to go too far upriver towards Glasgow; the three main

islands of the outer Clyde are conveniently at hand. Nearest the shore lies **Great Cumbrae**, a little island with a grand name, with the **Isle of Bute**, ten times its size, looming to the west. Both these islands were great favourites with the holidaymaking Glasgow crowds who'd throng 'doon the watter' on board the Clyde Puffers, crowded steamboats that plied the western Clyde. In the very outermost part of the firth lies **Arran**, tall and imposing, 'Scotland in miniature'. Arran is topped by the 2,868 ft (874 m) mountain of Goat Fell, one of three granite peaks that dominate an island whose dolerite, sandstone, granite and chalk have been squeezed, squashed and baked by volcanic upheavals into a spectacular geological mess that endlessly fascinates geologists.

## Inner Hebrides

Hoist your topsails and sheet home; you're bound north through what many consider the most romantic archipelago in the world, the Isles of the Inner Hebrides, mostly formed by spectacular volcanic explosions and outpourings. First on the horizon, as you round the Mull of Kintyre once more, are the southern twins of **Islay** (ragged peninsulas, fabulous malt whisky) and **Jura** (thirty times more red deer than humans, wild west coast of raised pebble beaches, quartzite peaks of the Paps of Jura, George Orwell's home while he wrote *1984*). Sail round the top of Jura, westward through the famed and feared Corryvreckan whirlpool, and with the Isle of **Colonsay** off your port quarter make north for the big, sprawling basalt layer-cake of the Isle of **Mull** (eagles and deer, retired English colonels, space and solitude). Memorable islands swim in Mull's shadow: **Iona**, cradle of Celtic Christianity, off the south-west tip; lumpy **Ulva** tucked into the west flank; further out, **Staffa** with its wonderful basalt columns forming the cathedral-like Fingal's Cave. Head north-west past the bizarre black sea-monster shapes of the **Treshnish Isles** to touch base with another pair of twins, **Coll** (pale rocky hills, corncrakes calling in traditional hayfields) and **Tiree** (windy, flowery, green, wave-pounded). Then haul back and north again through the 'Cocktail Isles' of **Rhum** (tall mountains, red deer, sea eagles), **Eigg** (lofty basalt cliffs, singing sands, a cave where the Macleods of Harris once massacred 395 Macdonalds – the entire population), and tiny, low-lying, carefully farmed **Muck**. Further north **Raasay**, an intriguing jumble of sandstone, ironstone and ancient gneiss, shelters between the mainland and the biggest, best-known and most northerly Inner Hebridean isle, **Skye**. Of romantic name and fame, Bonnie Prince Charlie's refuge with Flora Macdonald, Skye is a beautiful spread of five peninsulas with deep sea lochs and the crowning glory of the jagged Cuillin Hills.

## Outer Hebrides

Make a circuit of the coast of Skye; then bid farewell to the Misty Isle and set course to the south-west, out towards the Atlantic to pick up the tail of the great 130-mile (200-km) curve of islands that make up the Outer Hebrides archipelago (often referred to as the Western Isles). These rugged islands take the full impact of Atlantic weather on their ancient rocks, mostly 3,000-million-year-old Lewisian gneiss, the oldest rock in Britain. The southern end of the island chain is marked by the isle of **Berneray** (one of a number of islands with that name – it means 'Bjorn's island'; Bjorn must have got about a bit). Turning north, you cruise in and out of a whole string of small, steep green islands, deserted since the Highland Clearances of the nineteenth century – **Mingulay** (made famous by Sir Hugh Robertson's 'Mingulay Boat Song', a school choir staple: 'Hill you ho, boys, let her go, boys, sailing home, home to Mingulay'), **Pabbay** and **Sandray**. Next comes **Vatersay**, almost cut in two, and the roundish, hilly island of **Barra**, whose air passengers land on a superb beach of white cockleshell sand; then tiny **Eriskay**, where the wartime wreck of the freighter SS *Politician*, carrying £1 million in Jamaican banknotes and a quarter of a million bottles of prime whisky, sparked a year-long party and a famous novel by Barra resident Sir Compton Mackenzie – *Whisky Galore*. Next in line are the two Uists: **South Uist** (fabulous flowery beaches on the west, rearing mountains in the east), and **North Uist** (ditto), separated by the very flat and boggy **Benbecula**. More water than land, island of a thousand lochs, Benbecula is totally devoid of cover. Yet a grizzly bear named Hercules, eight feet tall and half a ton in weight, somehow managed to get himself lost there for three weeks in 1980 (he was starring in a Kleenex advert, naturally).

The top of the Outer Hebridean chain is composed of one single island – or is it two? **Lewis** and **Harris** are part of the same block of land – mountainous Harris to the south of huge, triangular, bog-and-loch-spattered Lewis; but for some reason too odd to figure out, they're referred to as two distinct isles. Whatever! These are the classic Western isles: harsh, bleak, Presbyterian and very, very windy. A plan to build more than 500 wind turbines, each 450 ft (137 m) tall, in the bogs of Lewis was turned down in 2009 after furious representations from ecologists, ornithologists and lovers of truly wild landscape – but watch that empty space . . .

## Far out

Fifty miles (80 km) west of the Western Isles, way out into the Atlantic, lies the archipelago of **St Kilda**. This is the ultimate goal for island-lovers, its main island a green bowl enclosing a deserted village, behind which the land steepens to enormously high cliffs. The other islands are even more dramatic and hard of access, pinnacles rearing from the sea, surrounded by swirling clouds of seabirds. The St Kildans, a tiny, hardy population, clung on in poverty and isolation, scaling the cliffs to harvest seabirds and their eggs, fishing and sheep-tending, until the difficulties of maintaining such a remote life in the modern world forced them to accept evacuation to mainland Scotland in 1930. But even St Kilda is not the remotest British island – not by a long chalk. You'd have a long day's sail west again (187⅓ miles/300 km), to be precise) to heave to under the lee of **Rockall**. Rocky, as the name suggests, sheer, bare, bleak, never inhabited (although very occasionally landed upon), and utterly isolated in the Atlantic Ocean, Rockall is as tall as the tail of a Boeing 747, and covers the same area as the foretopsail of HMS *Victory* – in other words, it's a tiny rock in a great big sea. It is nevertheless a piece of home ground. Rockall was formally annexed by the UK in 1972, and ex-SAS man Tom McClean lived on the rock for nine days in 1985 to reinforce the claim. The issue of ownership is not at all a light-hearted jape, but relates to the right to fish, drill for oil and gas, and search for ocean-floor minerals in the area. So other local states – notably Ireland, Denmark and Iceland – vigorously dispute Britain's sovereignty over this weather-beaten lump of volcanic detritus out in the middle of nowhere. It's a long haul north-east from Rockall back towards the Western Isles, giving a wide berth to haunted **Flannan Isle**, whose three lighthouse keepers disappeared without trace ten days before Christmas 1900, a never-solved mystery. Skirt the Butt of Lewis and Cape Wrath, steering east with a pinch of north to bump across the tide rips of the Pentland Firth (wave to starboard at the folks on the cliffs of Duncansby Head) and enter the wild seas of the Northern Isles.

## Northern Isles

Two archipelagos with very different characters make up the Northern Isles – the round sandstone cluster of the Orkney Islands, just off the northern coast of the Scottish mainland, and the long, ragged splatter of the granite Shetland Islands 50 miles (80 km) further north. Orkney Mainland is the chief island of the Orkneys, but before reaching it you turn west to view the mighty cliffs of the potato-shaped island of **Hoy**, where the great red sandstone **sea stack** (free-standing stone column) called the Old Man of Hoy, as tall as the London Eye, rises off shore. **Mainland**, a 30-mile (48-km)-wide splotch of peninsulas, bays and headlands, holds the capital of the archipelago, Kirkwall, and the 'second city', stone-built Stromness, and also boasts wonderful ancient structures, including Maes Howe, the finest chambered tomb in Europe; Skara Brae, a 4,000-year-old village of virtually untouched houses preserved in a sand dune; the far older stone circles, huge and mysterious, known as the Stones of Stenness and the Ring of Brodgar. From Mainland you wander slowly north by way of the isles of **Shapinsay** (a bizarre Victorian castle) and **Rousay** (the twenty-five-seater Stone Age tomb called the 'Ship of Death'), sandy **Sanday** (white beaches, turquoise sea) and the strong farming and fishing community of **Westray**. Two tiny isles guard the northern edge of Orkney – **Papa Westray** (with its enormous seabird colony, and the twin dwellings at the Knap of Howar which are the oldest houses in Europe), and **North Ronaldsay**, out at the rim of the archipelago.

It's a long 50 miles (80 km) running north-east from Orkney to Shetland, so why not break the journey halfway at Shetland's most southerly outlier, lonely **Fair Isle**? Here twitchers can hope to spot a bar-tailed thingummybobbit, or some such. Smack in the middle of an empty sea, out of sight of any other land, and positioned right under the track of vast migratory routes, Fair Isle is a famous landfall for rare birds. It's famous for hospitality, too, and for music and dancing – and that's exactly what to expect as you sail on north to reach the main Shetland archipelago. This is a bare and treeless environment of peaty hills and of black rock inlets called 'geos', cutting into the islands so deeply that nowhere in the 70-mile (113-km)-long archipelago is more than an hour's walk from the sea. The North Sea lies to the east, the Atlantic to the west. As with Orkney, the chief island of Shetland is called **Mainland**. Tie up in the harbour of the grey stone capital, Lerwick, for a session of Shetland's famous fiddle music in The Lounge bar, before heading north up the east coast of Mainland through the off islands of **Whalsay** and **Out Skerries** (fishing, and not much else). Alternatively you could head south from Lerwick to round the southerly point of Sumburgh Head and cruise up Mainland's west coast, with the stepped profile of the bird island of **Foula** to port and **Papa Stour** (caves and sea stacks) to starboard. Either way you'll get your fill of black cliffs, white sands and seabird rocks. North of Mainland you pass the big ragged-edged Isle of **Yell**, all moorland and lonely coasts, with another bird island, little-visited **Fetlar**, tucked into its eastern armpit. After that the great northerly cruise, well over a thousand miles from the Isles of Scilly, concludes with a run up the bleak shores of **Unst**, Britain's most northerly island. At the northern end of Unst you pass the gull-whitened sea stacks of Vesta Skerry, Rumblings, Tipta Skerry and lighthouse-crowned **Muckle Flugga** (wonderful names, full of sea echoes). Round to under the lee of the wave-smoothed round rock button of **Out Stack**. Look north and marvel – next stop, a thousand miles north, is the Arctic ice.

### Firth of Forth

From the Northern Isles you set your mind southward, a 600-mile (965-km) run with only a handful of islands to detain you. The contrast with the  island-spattered west coast of Scotland is quite remarkable. It's 250 miles (400 km) at least from Duncansby Head to the Firth of Forth, and there's not a single island worth turning aside for. Arriving off the wide mouth of the Firth of Forth you'll spy the Isle of **May**, famous as a breeding ground of puffins (up to 100,000 in a season). Bear west into the estuary, past the sizeable lump of **Inchkeith** (once an eccentric animal sanctuary, now uninhabited), and cruise between **Inchcolm** with its large, beautiful twelfth-century abbey and monastery buildings (built by King Alexander I in thanks for deliverance from a storm), and **Inchmickery**, whose profile was deliberately altered during the Second World War to look like a battleship. It still works – see for yourself! Round tiny **Inchgarvie**, right under the colossal span of the Forth Railway Bridge, and head on out again past little **Cramond** with its toast-rack causeway, on out to the morsels of **Fidra**, **Lamb** and **Craigleith**, and finally the huge granite outpost of the **Bass Rock**, where 100,000 gannets wheel and chakker eternally.

## Northumberland

Down across the imaginary line that delivers you from Scotland back into England, you find **Holy Island** at the end of its sandy causeway off the Northum-berland coast. Island of St Aidan and St Cuthbert, beacon of spirituality in the gloomy night of the Dark Ages, Holy Island (sometimes called by its beautiful name of Lindisfarne, 'retreat by the river') still exudes magic, especially by night when the seals are singing out on the sands. Within sight rise the submarine shapes of the **Farne Islands**, dark dolerite shelves canted in the sea; they represent the final hoorah of the volcanic Whin Sill (last seen in Upper Teesdale – see p. 114) before it dives into the North Sea. Guillemots, arctic skuas (who will savagely dive-bomb and peck you during the breeding season), shags and kittiwakes abound; so do grey seals, and local folk memories of young Grace Darling, daughter of the lighthouse keeper on the remotest island, Longstone, who rowed with her father on 7 September 1838 to rescue victims of a shipwreck on the Farnes. A few miles south and you'll be passing **Coquet Island**, another bird reserve island famous for its nesting terns. And that's pretty much the last proper island you'll sight for the next 250 miles (400 km), all the way to . . .

## Essex

Sea-stained and hungry for harbour, you round the eastward bulge of East Anglia and begin to stand in towards the River Thames once more. But before you smack your lips in anticipation of the voyage's end, cast an eye over the flat coasts and marshy, muddy islands of Essex on your starboard hand. This is some of the youngest land in Britain, made up of London clay laid down after the dinosaurs died out, shell-sand rich in iron, and a top dressing of mud and gravel left behind by the Ice Age glaciers only 10,000 years ago – the blink of an eye compared with that 3,000-million-year-old gneiss up in the Outer Hebrides at the opposite edge of Britain. Here in the throat of the River Blackwater is oval-shaped **Mersea Island**, where oysters are still sold fresh on the shell, and beyond it **Osea Island** (once a drying-out sanatorium for well-heeled drunks, later a motor torpedo boat base) and marshy **Northey Island** where in AD 991 a force of marauding Danes smashed an Anglo-Saxon army with terrible slaughter at the Battle of Maldon. And here at the mouth of the Thames is **Foulness** once more, with dead-flat **Potton Island** and **Wallasea Island** crouched in behind it. Amid so much manmade land, reclaimed from the sea and jealously guarded against it with such labour and care, Wallasea is where history is going into reverse and ushering in the future; because here the flood banks have been deliberately broken to let the sea back in, an experiment to see how much of a natural barrier of salt marsh will establish itself, and an acknowledgement that the level of the sea is rising and that we can't hold it back. Will the east coast be flooded in fifty years, or five hundred? If so, how extensively? We don't know. But the Wallasea Island experiment is the start of finding out – something to muse on as you sail proudly upriver, make fast at Wapping New Stairs, and stagger ashore for a well-earned pint in the Prospect of Whitby.

# 9

## LANDMARKS

Here are 100 great landmarks of the British Isles – features either natural or manmade, which chart our journeys like a series of beacons.

Sat Nav won't tell you about most of these. Some of them are so old it makes your eyes water just thinking about it; some are very recent additions to the scene. Some are right beside, or easily visible from, our main motorway and railway routes. Others you'll have to seek out, or even approach on foot. But each one is an old-school navigation aid, the kind of feature that catches your eye and makes you say, 'Oh, yes, I recognise that – I must be just *here*.'

I've chosen landmarks that are like old friends to me. But there'll be hundreds more I haven't thought of – landmarks that are your own personal favourites. Add them to the list as you go!

And if anyone can tell me (www.christophersomerville.co.uk) the name of that fabulous Victorian-looking building in red and black brick with fairy-tale turrets and gables, the one that looms over the M5 where it curves through the Black Country between Junctions 2 and 1, I'll be very, very grateful!

M = manmade  N = natural feature  ☐ = insert your own favourite landmark

## West Country

**1.** N *Dartmoor Tors, South Devon*
*(A30 near Okehampton, A38 near Bovey Tracey)*

Characteristic granite tors, hilltop outcrops that look like jagged castles. You know you're near dark and moody Dartmoor when you spot these on the skyline.

**2.** N *Lundy, North Devon*
*(Off A39 near Hartland; off B3343 near Woolacombe)*
The 3-mile (5-km)-long granite bar of Lundy, a dozen miles west of the North Devon coast, seen on the sea horizon on clear days, is everyone's idea of a romantic, sea-girt island.

**3.** M *Willow Man, Somerset*
*(M5 near Bridgwater)*

Willow Man, the creation of artist Serena de la Hey, strides out of the misty plains of the Somerset Levels next to the M5. Forty feet

(12 m) tall, his energy and strangeness catch the atmosphere of the peat moorlands.

### 4. M *Glastonbury Festival's 'Girt Wall', Somerset*
*(Worthy Farm, near Pilton)*
Now you see it, now you don't – the Girt Wall of Glastonbury. Erected every June, it guards the fields of Worthy Farm for the long celebratory weekend of the Glastonbury Festival. Then it's dismantled, and Michael Eavis's cows take over once again.

### 5. M/N *Glastonbury Tor, Somerset*
*(A39 near Glastonbury)*
The Tor itself, a hummock in the flat Levels landscape, isn't manmade. But would we pick it out so easily if it weren't for the medieval tower – all that remains of St Michael's Church – that crowns it so satisfyingly?

### 6. M *Clifton Suspension Bridge, Bristol*
*(A4 Portway, Bristol)*
Isambard Kingdom Brunel (1806–59), Britain's great Victorian engineer, left many marks in the British landscape, but this is the most iconic – his graceful suspension bridge, symbol of Bristol, that spans the Avon Gorge.

### 7. M *Severn Bridges*
*(M48 and M4, west of Bristol)*
The two Severn Bridges cross the wide Severn estuary a couple of miles apart. The older, upstream crossing, opened in 1966, carries the M48 between Bristol and Chepstow; the second bridge, opened in 1996, carries the M4 between Bristol and South Wales.

☐

☐

☐

## South Country

**8.** M/N *Isle of Portland, Dorset*
*(A354 at Weymouth)*
The 'Gibraltar of Wessex', a giant block of freestone quarried since Roman times, lies out in the sea like a sleeping lion.

**9.** N *Lulworth Cove, Dorset*
*(B3070 from Wareham)*
The sea has broken through a weak point in the hard freestone cliffs and burrowed out a spectacular, perfectly semicircular cove in the soft chalk and clay behind. One mile west, the sea has pierced a freestone promontory to make the tall arch of Durdle Door.

**10.** M *Maiden Castle, Dorset*
*(A35, A354 south of Dorchester)*
Britain's finest hill fort, visible for miles, created over 4,000 years from the Stone Age to the Iron Age. Multiple ramparts encircle the hilltop.

**11.** M *Cerne Abbas Giant, Dorset*
*(A352 between Sherborne and Dorchester)*

This 2,000-year-old chalk giant, 180 ft (55 m) tall, stands in priapic splendour on the hillside. He wields a great club above his head, and he's also very obviously 'pleased to see you'.

**12.** M *Stonehenge, Wiltshire*
*(Off A303 between Amesbury and Winterbourne Stoke)*
Europe's most famous stone 'henge' or monument, a mighty Bronze Age double circle of trilithons or stone gateways,

constructed on Salisbury Plain over a period of a thousand years. It may be a place of worship, a star computer or a giant calendar.

### 13. M *Silbury Hill, Wiltshire*
*(A4 just south of Avebury)*
Stone Age builders heaped up this flat-topped cone 5,000 years ago as a great hill of white chalk 120 ft (37 m) tall, with a radiating wheel of stone walls at its heart.

### 14. N *The Needles, Isle of Wight*
*(B3322 from Totland; also visible from B3058 in Milford-on-Sea on the mainland)*
Three huge blades of chalk,
dazzling white in sunshine, rise off the western tip of the Isle of Wight. Best aspect is from the viewing platform at the edge of the cliffs.

### 15. M *Spinnaker Tower, Portsmouth, Hampshire*
*(Portsmouth waterfront)*
Its graceful white sail shape visible for many miles along the south coast, the 558-ft (170-m)-tall Spinnaker Tower was opened in 2005 – a rather delayed millennium project.

### 16. M *Jack and Jill Windmills, Clayton, West Sussex*
*(A273 between Brighton and Burgess Hill)*
These two neighbouring windmills stand tall on their hill crest. Jill (built 1821) is a white-painted post mill; Jack (1866), a black tower mill. Jill was originally erected in Brighton, but was moved here in 1852. She still grinds organic flour from time to time.

### 17. M *Belle Tout Lighthouse, East Sussex*
*(On coast path between Beachy Head and Birling Gap)*
Standing proud on Belle Tout cliff, the lighthouse was built in 1834 and became redundant in 1902 when Beachy Head lighthouse superseded it. A remarkable operation in March 1999 saw the entire tower moved 50 ft (15 m) inland to prevent it falling over the fast-eroding cliff.

### 18. N *Beachy Head, East Sussex*
*(Off A259 between Newhaven and Eastbourne)*

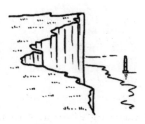

Beachy Head is the tallest chalk cliff in Britain, a mighty 530 ft (161.5 m) from its sheer (and unguarded) edge to the foot of its sloping, fan-shaped rampart. It dwarfs the lighthouse below – and that's almost as tall as Nelson's Column.

### 19. M *Little Cheyne Court Wind Farm, Romney Marsh, Kent*
*(Off A259 between Rye and Brookland)*
You can't help but have your eyes drawn to this giant development of twenty-six wind turbines, each as tall as the tallest redwood tree, whirling their arms round in a dead-flat landscape.

### 20. M *White Horse, Folkestone, Kent*
*(M20, Junction 12)*
Artist Charles Newington's millennium project, the Folkestone White Horse is cut into the slope of Cheriton Hill near the Eurotunnel terminal. Its vigorously cavorting shape is uplifting and inspirational.

### 21. M *Dover Castle, Kent*
*(Port and waterfront of Dover)*

Commanding the port of Dover and the whole seafront, Dover Castle squats like a great defensive beast on its high hilltop, a position it has dominated for nearly a thousand years.

### 22. N *White Cliffs of Dover, Kent*
*(From Dover's Western Docks, Folkestone Pier, or – best of all – a cross-Channel ferry)*

Landmarks don't come more iconic than this. The White Cliffs of Dover symbolise everything we like to believe of ourselves – strong, distinctive, upright and defiant.

☐
☐
☐

## Home Counties

### 23. M *London Eye*
*(From Westminster Bridge)*

The largest ferris wheel in Europe, towering beyond Big Ben and the Houses of Parliament, the London Eye is a landmark that's great to look at, but even better to look *from*.

### 24. M *Orbit Tower, Stratford, E. London*
*(Olympic Games 2012 site)*

A truly breathtaking concept from one of Britain's most controversial artists, Anish Kapoor's twisty Orbit Tower 'looks again at the whole idea of a tower, with an element of instability'.

### 25. M *Stokenchurch BT tower, Bucks*
*(M40, Junction 5)*

A reinforced concrete column, the Stokenchurch tower was built in 1968 as part of the 'Backbone' chain of microwave towers, which would maintain long-distance communications in the event of a nuclear attack wiping out the conventional telephone network.

### 26. M *Windsor Castle, Berks*
*(M4, between Junctions 6 and 5)*

The familiar silhouette of Windsor Castle, with its high walls and great Round Tower, dominates the skyline south of the M4 near Slough.

### 27. N *Devil's Punchbowl, Surrey*
*(A3 at Hindhead)*

Geologists say that the deep, steep-sided hollow of the Devil's Punchbowl was formed by springs washing away the underlying Lower Greensand and the clay below. Romantics maintain it was caused by the Devil when he gouged out a handful of rock to throw at the god Thor during one of their frequent play-fights.

☐

☐

☐

## East Anglia

### 28. M *Copped Hall, Essex*
*(M25 between Junctions 26 and 27)*

Everyone who drives on the M25 has seen this sinister shell of a Georgian mansion standing on high ground to the north of the motorway. Burned out in a fire in 1917, Copped Hall house, its gardens and parkland have all been saved from 'development', and currently await restoration.

### 29. M *Orford Ness 'pagodas', Suffolk*
*(B1084 from Woodbridge)*
One of the Cold War's numerous 'Secrets and Lies' sites around the UK, the pagoda-shaped structures on the lonely shingle spit of Orford Ness were laboratories where atomic bombs were shaken, heated, frozen, spun and banged on the floor to test their reactions . . .

### 30. M *Aldeburgh Martello Tower, Suffolk*
*(B1122, Aldeburgh seafront)*

Largest and most northerly of a string of 105 defensive towers built in 1805–12 to guard against a feared invasion by Napoleon Bonaparte, Aldeburgh's Martello Tower has so far proved defiant against the sea's best attempts to wash it away.

### 31. M *Bedford Rivers, Cambridgeshire*
*(A142 at Mepal, A1101 at Welney, A1122 at Denver Sluice)*
Two remarkable manmade rivers, 20 miles (32 km) long and ruler-straight, dug in the seventeenth century in a grand – and successful – attempt to stop the Fens flooding and to drain them for agriculture.

### 32. M/N *The Fens, Cambridgeshire/Norfolk/Lincolnshire*
*(Off A47 between Peterborough and King's Lynn; A141 between Huntingdon and March; A10 between Ely and Downham Market)*

You couldn't be anywhere else but the Fens – rich, powdery, black peat soil, dead-straight roads and rivers running high above the level of the land, lonely farms in hedgeless prairies of corn and vegetables, all under a huge open sky.

### 33. M *Boston Stump, Lincolnshire*
*(Boston town centre; roads for 30 miles [50 km] around)*
Known affectionately as 'Boston Stump', the 272-ft (83-m) tower of St Botolph's Church by the River Witham is a landmark and beacon over a huge Fenland countryside. There's a stupendous prospect from its viewing platform.

### 34. M *Alkborough Flats, Lincolnshire*
*(Minor road in Alkborough, off A1077 between Scunthorpe and Barton-upon-Humber)*
The view from the ancient grass maze of Julian's Bower (signposted) shows you a mighty landscape where the rivers Trent and Ouse join to form the River Humber – the scene of a landmark experiment in sea level control, where the sea walls around Alkborough Flats at the foot of the hill have been deliberately breached to allow tidal flooding and the creation of new mudflats and saltmarsh.

### 35. M *Humber Bridge, Lincolnshire/East Yorkshire*
*(A15 near Hull)*
The east of England's greatest bridge, a span nearly 3 miles (4.8 km) from bank to bank of the River Humber, which you can walk across if you have a good head for heights.
☐
☐
☐

## South Midlands

**36.** M *National Lift Tower, Northampton*
*(Off A45, near M1 Junction 15a)*
A wonderful piece of engineering that's as much a work of art as it is a functional building, the National Lift Tower was opened in 1982 to test the elevators of the Express Lift Company. It's 418 ft (127 m) tall; slender, tapering and beautiful.

**37.** M *Uffington White Horse, Oxfordshire*
*(From B4507 and minor roads north around Uffington)*
One of the UK's oldest manmade landmarks, this exceptionally expressive chalk figure of a wildly  galloping, disjointed horse is so minimalist it could easily be a modern artwork, but in fact it was probably cut around three thousand years ago.

**38.** M *Edgehill Castle, Warwickshire*
*(Off M40 Junctions 11, 12; from minor roads around Radway, between B4086 Kineton–Warmington and A422 Ettington–Banbury)*
Splendidly set on the skyline of a great ridge, this Gothic tower (now an inn) commands an enormous view over the English Civil War battlefield of Edgehill (1642 – a draw between Parliament and the King), whose centenary it was built to commemorate.

**39.** N *Malvern Hills*
*(M5, Junctions 7–11)*
The dinosaur spine of the Malvern Hills is a landmark for any M5 travellers who glance west between Gloucester and Worcester. Capped with granites 1,000 million years old, the

Malverns look like a mountain chain – although a good walker can nip up to the top in a few minutes.

**40.** M *RAC Centre, Bescot, Birmingham*
*(M6, Junction 8)*
Designed by an Italian architect (can't you tell?), the RAC's stylish Bescot operations control centre has reared its glass wall and sloping roof over the M6/M5 junction since 1989.

☐

☐

☐

## North Midlands

**41.** M *Rutland Water, Rutland*
*(From A606 Oakham–Stamford, around Whitwell; south shore cycle path around Edith Weston; Upper Hambleton on the peninsula)*
This lovely lake, hugely popular as a day-out destination, is actually a reservoir, built in the 1970s to supply the East Midlands with water.

**42.** M/N *Breedon-on-the-Hill cliff, Leicestershire*
*(From A453 between M1 Junction 24 and Ashby-de-la-Zouch)*
A memorable sight – the twelfth-century priory church of Breedon-on-the-Hill apparently teeters on the very brink of a tall limestone cliff. The church contains wonderful Dark Ages carvings, by the way.

**43.** M *Clipstone Colliery headstocks, Nottinghamshire*
*(B6030 between Mansfield and Ollerton)*
Taller than the tallest trees in Europe, the giant twin head-stocks of the former Clipstone Colliery (closed 2003) tower

over the site. Not for much longer, perhaps – locals have voted for the demolition of the 215-ft (66-m)-tall headstocks, grim reminders to them of too much sweat and blood spilt in the recent past.

### 44. M *Crooked spire, Chesterfield, Derbyshire*
*(From A61 or railway through town)*

The crooked spire of Chesterfield's parish church, added about AD 1362, is a well-known landmark for road and railway travellers. The medieval architects probably intended to give it a barley-sugar twist, but its lean (almost 10 ft [3 m] out of true!) resulted from the use of unseasoned wood and the omission of bracing timbers by greenhorn builders – most experienced tradesmen had died during the Black Death epidemic a dozen years before.

### 45. N *Stanage Edge, Derbyshire*
*(Minor road off A6187 north of Hathersage)*

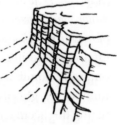

A striking gritstone cliff several miles long; its face, weathered into pinnacles, ledges and cracks, is perfect for rock climbing and bouldering, while its flat top is used as a launching pad by hang-gliders and paragliders.

### 46. M *Derwent Dam, Derwent Reservoir, Derbyshire*
*(A57 Sheffield–Manchester at Ladybower Reservoir; minor road [brown 'Derwent Valley' sign] for 2¼ miles [3.6 km] to visitor car park; view dam from 'No Vehicles' road beyond Visitor Centre)*

There's no more iconic dam in Britain: this was the target of RAF Bomber Command's 617 Squadron during weeks of intensive practice in 1943, before the 'Dambusters' flew their Lancaster bombers to destroy the Möhne and Eder dams with Barnes Wallis's revolutionary 'bouncing bombs'.

### 47. N *Dovedale, Derbyshire*
*(Footpath from Dovedale car park near Thorpe, off A515 north of Ashbourne)*
Remarkable gash in the limestone of Derbyshire's White Peak district, a favourite among walkers with its lofty pinnacles of the Twelve Apostles, Tissington Spires, Ilam Rock and Pickering Tor.

☐
☐
☐

## Welsh Borders

### 48. N *Stiperstones, Shropshire*
*(Stiperstones footpath from car park – OS ref SJ 369977 – on minor road between Bridges and The Bog field centre, east of A488 Shrewsbury–Bishop's Castle)*
Ominous and ill-omened, this line of legend-laden quartzite outcrops dominates the skyline. The Devil is only one of a number of ghoulies, ghosties and long-legged beasties to haunt the accursed stones.

### 49. M *Ironbridge, Shropshire*
*(From riverside footpath, off A4169)*
Spanning a gorge on the River Severn is the classic landmark of the early Industrial Revolution. Abraham Darby's pioneering bridge of 1779 was the world's first cast-iron

arch bridge, making a perfect circle with its own reflection in the river.

**50.** N *The Wrekin, Shropshire*
*(M54, Junctions 5–7; A4169*
*Ironbridge–Wellington)*

The handsome ridge of the Wrekin, a mini-mountain shaped like a shallow-sided tent, draws the eye in an undramatic part of the world.

☐
☐
☐

## Yorkshire

**51.** M *Ferrybridge power station cooling towers, West Yorkshire*
*(At junction of M62 and A1)*

The eight cooling towers at Ferrybridge power station dribble clouds of water vapour like a group of amiably smoking giants as they stand beside the junction of the M62 and A1, a familiar landmark to drivers, their truncated cones flaring gently out towards the top. You could drop a Big Ben inside each one, and still have room for a Tyrannosaurus Rex on top.

**52.** M *Emley Moor Tower, West Yorkshire*
*(From M1 around Junction 38; on Kirkburton–Emley minor road between A629 and A636)*

The slender concrete tower on Emley Moor, topped by its lattice mast, soars a mighty 1,084 ft (330 m). You'd need three of the Ferrybridge cooling towers, stacked one on top of another, to accommodate this telecommunications and broadcasting station. It's the tallest freestanding structure in these islands.

**53.** M *Stoodley Pike monument, West Yorkshire*
*(A646 Todmorden–Hebden Bridge; Pennine Way National Trail runs past it)*
Splendidly perched at the 'elbow' of its Pennine ridge, this stout stone obelisk commemorates two events: Napoleon Bonaparte's ultimate defeat at the Battle of Waterloo in 1815, and the end of the Crimean War in 1856.

**54.** N *Gordale Scar, North Yorkshire*
*(Signposted from Malham, off A65 between Settle and Skipton)*
A landmark for walkers rather than drivers, only those on foot will get to marvel at this extraordinary deep gash in the limestone moors of the Yorkshire Dales, a canyon carved out by a rush of Ice Age meltwaters.

**55.** M *Ribblehead Viaduct, North Yorkshire*
*(Moorland track off B6255 Ingleton–Hawes, near junction with B6479 Horton-in-Ribblesdale road)*

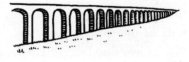

A breathtaking view – the most iconic railway viaduct in Britain, striding north on twenty-four tall, slender arches. Opened in 1875, it carries the Settle & Carlisle line across a lonely valley, with the long shoulder of Whernside rising beyond – an unofficial monument to the dozens of navvies who died building this remote line.

**56.** N *Hole of Horcum, North Yorkshire*
*(Car park viewpoint on A169 Pickering–Whitby, or walk from Newton Dale station on North York Moors Railway)*
A vast hollow 400 feet (120 m) deep and ¾ mile (1.2 km)

wide, formed by glacial meltwater – or maybe by the Devil when he scooped out a clawful of the moor to fling at a witch who had cheated him.

### 57. N *Roseberry Topping, North Yorkshire*
*(From A173 between Great Ayton and Guisborough)*

Not an instant pudding, as you might think from the name, but a landmark hill, crested like a breaking wave. In 1914 half the hill fell away in a landslip that produced its unmistakeable shape.

- ☐
- ☐
- ☐

## North-West

### 58. M *Liverpool Pierhead and Liver Birds, Merseyside*
*(Liverpool waterfront)*

Three magnificent buildings characterise Liverpool's Pierhead: the Port of Liverpool Building under its dome, the square bulk of the Cunard Building, and the massive towers of the Royal Liver Building, where the city's iconic Liver Birds perch.

### 59. M *'Another Place', Crosby*
*(Signposted from Crosby on A5036)*

Installed on Crosby Beach in 2005, the 100 naked, cast-iron men that make up sculptor Antony Gormley's 'Another Place' have courted controversy. Some locals want them removed as obstructive, dangerous or even obscene. Others find them beautiful, stirring and challenging. Now the dust has settled, it looks as though they're there to stay. Huzzah!

### 60. M *Blackpool Tower, Lancashire*
*(Blackpool seafront)*

This absolute and ultimate symbol of Victorian seaside Britain, visible for many a mile along the Lancashire coast, was built in 1894 to sprinkle over the working-class resort of Blackpool some of the glamour associated with Paris and the kind-of-similar-but-taller-and-classier Eiffel Tower.

### 61. M *Ashton Memorial, Williamson Park, Lancaster*
*(From M6 at Lancaster, between Junctions 33 and 34)*

Q: What's that extraordinary-looking wedding-cake building topped with a green dome that emerges from the treetops beside the M6 at Lancaster?

A: It's the (possibly haunted) Ashton Memorial in Williamson Park, and it was built in 1907–9 by the 'Oilcloth King', millionaire James Williamson, First Baron Ashton, a local boy, in memory of his second wife Jessy.

### 62. N *Howgill Fells, Cumbria*
*(From M6, between Junctions 37 and 38)*

Q: What are those dramatically folded hills, rounded and green, that rise so enticingly on the east of the M6 just north of Killington Lake services?

A: They're the Howgill Fells, made of a hard sandstone called Coniston Grit that resists weathering and forms rounded summits. Fabulous walking here, and the dramatic Cautley Spout waterfall is tucked away round the back of the range.

### 63. N *Striding Edge, Cumbria*

*(Walk up from Glenridding or Patterdale on A592)*

Scary, head-for-heights teeter towards the summit of 3,118-ft (950-m) Helvellyn along a razorback ridge one foot wide, with slopes plunging hundreds of feet on either side. Every hill walker's baptism of fire. Wonderful views from Nethermost Pike, from Helvellyn summit, and of course from the start of the Edge itself.

☐

☐

☐

## North-East

### 64. N *High Force, Co. Durham*

*(Footpath opposite High Force Hotel on B6277 Middleton–Alston)*

You can't see it from the road, but descend the footpath to the foot of High Force and you'll see why this heavyweight bruiser of a cataract, 70 feet (20 m) of thunderous water crashing down the black rock step of the Whin Sill, is a landmark waterfall.

### 65. M *Tees Transporter Bridge, Teesside*

*(On Durham Street/Ferry Road)*

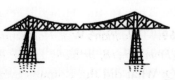

Middlesbrough's trademark structure, the Tees Transporter Bridge carries vehicles and people in a gondola, 160 feet (50 m) above road level, across the River Tees. Built before the First World War, its skeletal animal shape has endeared it to locals.

## 66. M *Durham Cathedral and Castle, Co. Durham*
*(Best view from Durham's railway viaduct)*

The classic landmark railway view as you slide across the viaduct in the train – Durham's promontory standing high over its bend of the River Wear, with the castle and cathedral outlined in heavy, powerful splendour on top.

## 67. M *Angel of the North, Tyne & Wear*
*(On A167, where A1 splits from A194 on southern outskirts of Gateshead)*

Is it a bird, is it a plane, is it a man with his arms strapped to an ironing board? No, it's Antony 'Another Place' Gormley's fabulous sculpture, *Angel of the North*, with the height of a tall factory chimney and the wingspan of a Boeing 767F, keeping watch from its knoll.

## 68. M *Tyne Bridge, Tyne & Wear*
*(Newcastle city centre; excellent views from High Level railway bridge and Millennium pedestrian/cyclist bridge)*

Newcastle-upon-Tyne boasts several handsome bridges, but only one has national landmark status – the green bow of the Tyne Bridge (opened 1928), which carries the A167 across the River Tyne between Newcastle and Gateshead.

## 69. M *Hadrian's Wall, Northumberland*
*(From B6318, or along Hadrian's Wall Path National Trail)*

Q: What did the Romans ever do for us?

A: They left us the finest Roman monument in these islands, Hadrian's Wall (started AD 122), the mighty defensive structure that runs for 73 miles (117 km) right across the neck of the English/Scottish border from Newcastle-upon-Tyne to the Solway Firth beyond Carlisle.

### 70. N *Whin Sill, Northumberland*
*(From Hadrian's Wall; at Bamburgh Castle and the Farne Islands)*
The tongue of dark volcanic dolerite that intruded into the rocks of northern England some 300 million years ago solidified into a giant step, known as the Whin Sill (see 64 High Force), that outcrops spectacularly in Northumberland – as a line of inland cliffs that runs for miles topped by Hadrian's Wall, as the coastal ledge on which perches Bamburgh Castle, and as the bare black sea-monster shapes of the Farne Islands just offshore.

☐

☐

☐

## Scotland

### 71. N *Ailsa Craig*
*(From A77 near Girvan)*
Eight miles (13 km) out to sea, the volcanic plug of Ailsa Craig rises like a leviathan – it's three times the height of the giant Bass Rock in the Firth of Forth. Ailsa Craig's blue hone granite is harvested to make curling stones for Scotland's national ice sport.

### 72. M *Falkirk Wheel*
*(From B816 at Tamfourhill, west of Falkirk)*
Massive and impressive in the classic manner of British heavy 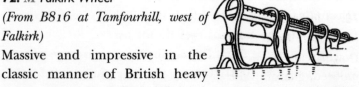 engineering, the Falkirk Wheel (opened 2002) looks like

the articulated skeleton of a dinosaur, but is a modern solution to a very old problem in connecting Glasgow and Edinburgh by water. This rotating boat lift and aqueduct brings vessels up and down the 80-ft (24-m) gap in level between the Union Canal and the Forth & Clyde Canal, doing away with the need for dozens of locks.

### 73. M *Holyrood building*
*(At foot of Royal Mile, Edinburgh)*

Designed by Catalan architect Enric Miralles (who said he wanted to make it 'grow out of the land'), the Scottish Parliament building (opened 2004) with its heaped-up appearance and futuristic windows and walls is proof positive that landmark buildings don't have to be old or long-established.

### 74. M *Forth Railway Bridge*
*(From B924 at Queensferry [south bank], B981 at North Queensferry [north bank] or Forth Road Bridge)*

The three skeleton cantilevers of the Forth Railway Bridge (opened 1890), as massive as the humps on a dinosaur's back, are painted an eye-catching red. This proud piece of Victorian engineering doesn't want to blend into the low-lying, green and brown landscape – it's all about Man's dominion over Nature.

### 75. M *Inchmickery*
*(From Forth Railway Bridge)*

Downstream of the Forth Railway Bridge a ship of war lies moored, guarding the naval dockyard of Rosyth. That's the way it looks – and has done ever since the Second World

War, when the upper slopes of the boat-shaped island of Inchmickery were built up with towers to resemble the superstructure of a battleship, all to fool German U-boats that might be contemplating an attack on Rosyth.

### 76. M *Stirling Castle*
*(From M9 near Junction 10)*
Half manmade fortress and half rock, Stirling Castle rides its high cliffs magnificently, a statement of power and majesty.

### 77. M *Bell Rock lighthouse, off Arbroath*
*(From A930, looking east-south-east), from A92 at Arbroath, looking south-east; or, close-to, on boat trip from Arbroath)*
Iconic lighthouse, built on the notorious Bell Rock reef in 1807–11 with huge difficulty in very dangerous conditions, visible in clear weather as a warning finger lifted on the sea horizon.

### 78. M *Rothes Windfarm, Moray*
*(From B9010 Elgin–Forres, or minor road between Dallas and Upper Knockando on B9102)*
Rothes Windfarm had twenty-two turbines 'not exceeding 100 metres [328 ft] in height' when it opened in 2005 on a lonely moor. Now there are plans to double the capacity and introduce taller turbines.

### 79. M *Skye Bridge, Highland*
*(A87 at Kyle of Lochalsh)*
Many objected when a bridge between the mainland and the Isle of Skye was mooted. Its opening in 1995 killed the

ferry service, and some of the romance, but there's no doubt the bridge is a landmark in the region.

**80.** N *Cuillin Hills*
*(From A87 and A893; close-to, minor road from B8009 at Merkadale to Glenbrittle)*
The Isle of Skye's iconic hills – the Red Cuillin of granite glows fox-red in the sun, and the Black Cuillin, of dark volcanic gabbro, forms classic peaks like a row of jagged teeth bared to the sky.

**81.** N *Treshnish Isles and Staffa*
*(Best seen from a boat from Fionnphort, Isle of Mull)*
Basalt's masterpiece in the Inner Hebrides. The molten volcanic rock cooled to produce extraordinary hexagonal columns (Staffa) and bizarre, submarine-like shapes (Treshnish). Once seen, never forgotten.

**82.** N *Suilven*
*(From minor roads around Lochinver at end of A837)*
Magnificent mountain, uniquely shaped, with a striated (horizontally lined) southern face, which resembles either a lion or a sphinx, depending on the weather and your mood.

**83.** N *Ben Loyal*
*(From minor road between Tongue on A838 and Kinloch Lodge)*
Not the biggest mountain in Scotland, but one of the most recognisable, its four rocky peaks rising on the western edge of the great peat-bog wilderness of the Flow Country.

**84.** N *Old Man of Hoy, Isle of Hoy, Orkney Isles*
*(From Scrabster–Stromness ferry)*
The 450-ft (137-m) rock stack of the Old Man
of Hoy rises off the Isle of Hoy like a crazed
sculptor's notion of a sandstone chimney. The
height of the Great Pyramid of Cheops, it is
quite astonishing that the sea-battered Old Man
hasn't yet fallen.

☐

☐

☐

## Wales

**85.** M *Port Talbot Steelworks, West Glamorgan*
*(M4, Junctions 38 and 39)*
You used to see their smoking stacks, furnace flares and giant's
geometry of pipework all over the country, but functioning
steelworks are rare birds these days. The Margam works by
the M4 in South Wales is one of the most eye-catching
anywhere.

**86.** N *Worm's Head, Gower, West Glamorgan*
*(B4247 to Rhossili; footpath to causeway*
*[for tide times, www.nciwormshead.org.uk*
*or phone: 01792 390707])*
You can't mistake this – a 2-mile (3.2
km)-long, green-backed promontory

of humps, sticking out from the western tip of the Gower
Peninsula like a monster putting out to sea. A thrilling, tide-
dependent scramble. Norsemen named it *wurm* ('dragon'),
and you can easily see why.

**87.** N *Pistyll Rhaeadr, Powys*
*(B4580 to Llanrhaeadr-ym-Mochnant; follow signposted single-track road)*
The highest waterfall in Wales, 240 ft (73 m) of tumbling water with a natural rock arch to jet through halfway down, and a pool of rainbows to crash into at the bottom.

**88.** M *Bersham Bank colliery tip, Clwyd*
*(Near Rhosllanerchrugog turning on A483 near Wrexham)*
An estimated six million tons of colliery waste form the massive, flat-topped slag mountain of the former Bersham Colliery (active 1879–1986); an eye-catcher (some would say eyesore) for many a mile.

**89.** M *Pontcysyllte Viaduct, Clwyd*
*(From B5434, bounded by A5 and A542 between Ruabon and Llangollen)*
The most beautiful aqueduct in Britain, designed by Thomas Telford and opened in 1805 to carry boats on the Llangollen Canal, it passes high over a deep valley of the River Dee on eighteen shapely brick arches.

**90.** M *Castell Dinas Bran, Clwyd*
*(From Llangollen)*
Rising far over the town on their dramatic 1,000-ft (305-m) knoll stand the ruins of ancient Castell Dinas Bran. Superb to look at from below, and stunning to look out from when you've climbed the steep path.

**91.** M *Dinorwig Slate Mines, Gwynedd*
*(From A4086 Caernarfon–Capel Curig)*
A remarkable vision of Hell, created by slate miners in the
mountain slopes of Mynydd Perfedd. Black craters, ledges,
levels, cliffs, screes and downfalls; a terrifying, ominous scene
of ruin in mist and rain.

**92.** M *Menai Bridges, Gwynedd*
*(A5 between Llanfairpwllgwyngyllgogerychwyrndrobwllllantysilio-*
*gogogoch and Menai Bridge)*
Splendid view of the two most famous bridges in North
Wales: Thomas Telford's suspension bridge of 1826 which
carries the A45 road, and Robert Stephenson's tubular
Britannia Bridge of 1850 which takes both the A55 and the
railway line to Holyhead and the Irish ferries.

**93.** M *North Hoyle Wind Farm, Clwyd*
*(A548 North Wales coast road around Rhyl and Prestatyn)*
Once there were thirty, now there are fifty-five giant
white turbines footed on the North Hoyle and Rhyl sandbanks 4 miles (6.5 km) offshore, their restless motion
along the northern sea horizon making them impossible to
ignore.
☐
☐
☐

## Northern Ireland

**94.** M *Belfast City Hall*
*(Donegall Square, Belfast city centre)*
The vast white wedding-cake edifice of Belfast City Hall

(opened 1906) epitomises a particular style of 'Imperial Pride' architecture, typical of Edwardian times.

## 95. M *Crumlin Road Gaol and Courthouse, Belfast*
*(Crumlin Road)*
A world away from the immaculately maintained City Hall are these (currently) burned-out and derelict twin monsters either side of the Crumlin Road, linked by a subterranean passage and their own grim histories.

## 96. M *Samson and Goliath*
*(From A2 and M2; also from Queen's Road)*
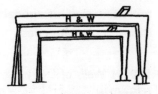
Iconic pieces of Belfast history, the two vast, yellow-painted cranes called Samson (started work 1974) and Goliath (1969) can be seen for miles beyond their home in Harland & Wolff's shipyard (where SS *Titanic* was built and launched in 1911).

## 97. M *Armagh Cathedrals*
*(B115 runs between the two cathedrals)*
Both cathedrals are called St Patrick's, both stand on knolls in the ancient city of Armagh. One – squat and medieval under its stumpy tower – is Church of Ireland; the other – large, flamboyant, ornate – is Roman Catholic. Facing each other across a narrow divide, they can't help but be symbols as well as landmarks.

**98.** N *Slemish, Co. Antrim*
*(From A42 Broughshane–Carnlough)*
The volcanic plug of a mountain where St Patrick, patron
saint of Ireland, herded pigs as a slave in his youth is a touchstone in Irish spirituality and a landmark for miles around
in the flattish mid-Antrim landscape.

**99.** N *Giant's Causeway, Co. Antrim*
*(Off B147 Causeway Road, near Bushmills on A2)*
Northern Ireland's best-known landmark, this great promontory of 37,000 hexagonal basalt columns stretches out north
from the Antrim coast.

**100.** M *Walls of Derry*
*(Derry city centre)*
Iconic to both Loyalists and Nationalists for very different
reasons, this tight ring of stone walls with a broad walkway
on top was built in 1613–18 by trade guilds from London
that had arrived to take over and develop the ancient Irish
settlement by the River Foyle.

☐
☐
☐

# 10

## WHAT DOES *THAT* MEAN?

### GEOGRAPHICAL TERMS

I was a clumsy cricketer at school (nothing has changed over the ensuing decades) – couldn't catch a ball, even if it was lobbed very slowly and gently right into my hands. *Plop!* Out again and down on the grass, as shiny and red as my cheeks. It was the same in the classroom, and especially in Mr Matt's Geography lessons. I just couldn't grasp those technical terms as they came flying by. It was partly the shape of the weird words, their exotic taste, their alien quality. In maths one heard of triangles, of pyramids and cones, but those were things already on one's radar – you made a donging sound with them in music, or saw men on camels riding past them in Biggles books, or licked your brick of chocolate ice-cream

out of them. But Geography was a whole new level of incomprehension. 'Moraine.' That didn't chime with anything else.

A bit like a migraine ... Terry French got those sometimes ... and a bit like moron, that was one of those words that wasn't exactly rude, but you got a sort of naughty kick out of muttering it ... By the time I'd stopped rolling moraine round my palate and hauled my attention back on track, oxbows and meanders and drumlins had all steamed right on by.

Here's a glossary of most of the technical terms you'll find in this book. You might know the lot; there might be a few you need to brush up on; or, like me back then, you might be struggling with a surprising number of them. Don't be shamefaced – God, we all have enough holes in our thin little socks of knowledge!

I've also suggested where to find a good example of each item.

**archipelago**: a chain or group of islands.
*Example*: The Outer Hebrides islands, Western Scotland.

**backwaters**: waters diverted by opposing currents.
*Example*: River Test around Stockbridge, Hampshire.

**bay**: an area of water that is partly surrounded by land.
*Example*: Cardigan Bay, West Wales.

**beck, gill, ghyll, burn, nant**: all regional terms for a stream – beck and gill/ghyll in the north of England, burn in Scotland and Northern Ireland, nant in Wales.
*Example*: Malham Beck, North Yorkshire; Ease Gill, Cumbria/ Lancashire border; Sourmilk Ghyll, near Grasmere, Lake District; Ben Glas Burn, Loch Lomond; Nant Fridd Fawr, Cadair Idris, West Wales.

**bog, bogland**: area of land where the underlying rock is so acid that the vegetation can't rot; instead it forms a thick

layer of peat, preserving seeds, timber, pollen and dead bodies. A **blanket bog** is a rain-fed bog in an upland area, which blankets the ground.
*Example*: Flow Country, northern Scotland.

**cataract**: big, powerful, plunging waterfall.
*Example*: Cautley Spout in the Howgill Fells, Cumbria, England's highest cataract at 580 ft (175 m).

**catchment**: the whole area drained by a river and its TRIBUTARIES.
*Example*: The River Medway's catchment is 930 square miles/2,409 sq. km – about two-thirds of the whole county of Kent.

**causeway**: land bridge (either manmade or natural), generally covered at high tide and exposed at low tide, connecting the mainland with an island or PROMONTORY.
*Example*: Worm's Head, Gower Peninsula, South Wales.

**cliff**: steep slope, sometimes a sheer drop, of land into the sea.
*Example*: Beachy Head in East Sussex is the highest chalk cliff in England at 530 ft (162 m).

**confluence**: where one river joins another.
*Example*: The River Humber between Lincolnshire and Yorkshire is formed by the confluence of two major rivers, the Yorkshire Ouse and the Trent.

**cove**: a small, BAY-shaped inlet on the coast.
*Example*: Kynance Cove, the Lizard Peninsula, Cornwall.

**creek**: a tidal sea inlet, generally on a low-lying, muddy coast.
*Example*: Brightlingsea Creek, Essex.

**cwm**: Welsh valley.
*Example*: Cwm Rhondda, the Rhondda Valley in South Wales.

**dale**: a valley – a Norse word in origin, often associated with the classic Yorkshire Dales pattern of grass pasture in the bottom, rougher land up the sides, and uncultivated moor at the top.
*Example*: Wharfedale, North Yorkshire.

**delta**: branching web of water channels.
*Example*: The Wash.

**downs, downland**: rolling, hilly landscape with chalk underlying it.
*Example*: The South Downs of Sussex.

**drumlin**: heap of gravel, stones, rubble left behind by a melting GLACIER, weathered into the shape of a smoothly rounded hill.
*Example*: South Armagh, near the Northern Ireland/Republic of Ireland border.

**dune**: sandhill heaped up by tides and sculpted by the wind.
*Example*: Sefton Coast of Lancashire.

**enriched**: *see* NITRATE ENRICHMENT.

**erratic**: boulder of an alien rock type, carried and deposited by a GLACIER.
*Example*: boulders of Shap GRANITE and millstone grit on the beach between Barmston and Bridlington, East Yorkshire coast.

**estuary**: where a river widens to meet the sea.
*Example*: Severn Estuary.

**fault**: a break in an area of rock where one or both surfaces of the fracture have moved some distance.
*Example*: The Great Glen, north-west Scotland.

**fell**: term used in the Lake District and nearby Pennine region for a hill.
*Example*: Bow Fell, near Scafell Pike, Lake District.

**fen**: freshwater wetland, usually with neutral or lime-bearing water – unlike BOGS, which are acid and fed by rainwater.
*Example*: Wicken Fen National Nature Reserve, Cambridgeshire.

**firth**: Scottish term for an ESTUARY.
*Example*: Firth of Forth, south-east Scotland.

**floodplain**: flat area often or occasionally flooded by a river (though sometimes protected by flood banks).
*Example*: There are several sections of floodplain along the River Severn, which regularly flood – for example, at Bewdley in Worcestershire, and between Tewkesbury and Gloucester in Gloucestershire.

**freshwater marsh**: *see* MARSH.

**glacier**: a mass of ice that moves slowly over the land, sometimes scratching deep grooves in the rocks it's travelling across.
*Example*: Glacial scratches in Coppermines Valley near Coniston, Lake District.

**glen**: Scottish word for a valley, generally higher up the river than a STRATH.
*Example*: Glen Nevis, western Scotland.

**gorge**: deep, narrow cleft cut into the landscape by a river. Some gorges are what's left of a water-burrowed cave after the roof has fallen in.
*Example*: Cheddar Gorge, Somerset.

**harbour**: strengthened refuge for fishing vessels and pleasure craft.
*Example*: Portland Harbour, Dorset.

**headland**: a point of land sticking out towards the sea, with water ahead and on both sides.
*Example*: Gurnards Head, Cornwall.

**heath**: a dry-ish habitat of heather, gorse and scrub, usually on acid, sandy soil.
*Example*: Dunwich Heath, Suffolk.

**islet**: small island.
*Example*: Midland Isle is an islet between Skomer Island and the Pembrokeshire mainland in West Wales.

**lake, loch, lough**: respectively the English, Scottish and Northern Ireland terms for a body of water surrounded by land. In the latter two countries, tidal inlets are known as sea lochs/sea loughs.
*Example*: Rudyard Lake, Staffordshire; Loch Lomond, Scotland; Lough Neagh, Northern Ireland. *Sea loch*: Loch Fyne, south-west Scotland. *Sea lough*: Strangford Lough, Co. Down, Northern Ireland.

**lakelet, lochan**: English/Scottish term for a small lake.
*Example*: There are dozens of lakelets between Northwich and Knutsford, Cheshire; the peat moors of the Isle of Lewis contain thousands of lochans.

**landlocked**: entirely surrounded by land.
*Example*: Wiltshire is a landlocked county.

**marsh**: wetland that is regularly or frequently flooded. A **saltmarsh** is flooded by the sea; a **freshwater marsh** tends to be a piece of land reclaimed from the sea and drained for grazing.
*Example*: At the western end of Canvey Island, Essex, horses graze the freshwater marshes that lie inside the sea wall, protected from tidal floods. At the eastern end, saltmarsh has developed outside the sea wall, between the wall and the low-water mark.

**meadow**: grassy field grazed by animals and harvested for hay.
*Example*: Somerset's cattle give good rich milk from grazing the county's meadows.

**meander**: the snaky bends formed by a river as it finds forward momentum through its flat FLOODPLAIN.
*Example*: River Wye below Hereford; Cuckmere River, Cuckmere Haven, East Sussex.

**meltwater**: the water that pours from a melting glacier, often with enough force to carve a way through rock.
*Example*: Meltwater created the cavern whose roof collapsed to form Gordale Scar in North Yorkshire.

**moor**: open, upland country, usually of heather and peat, on poorly drained acid soil.
*Example*: Dartmoor, Exmoor, North York Moors.

**moraine**: heap of rubble pushed by an advancing GLACIER; a **terminal moraine** is positioned at the furthest point that the glacier reached before starting to retreat.

*Example*: Coire an t-Sneachda ('corrie of the snows') above Glenmore Lodge in the Cairngorm Mountains. Visit: http://www.fettes.com/cairngorms/coire%20Sneachda.htm.

**mudflats**: as the name suggests, large areas of mud (sometimes gently domed, in fact) at the mouth of rivers/creeks, usually tidal.
*Example*: The Wash, Norfolk/Lincolnshire border; rivers Crouch, Roach, Blackwater in Essex.

**nitrate enrichment**: process by which nitrates from agricultural fertilisers and human/animal waste are washed by rain or drainage into rivers. The nitrates 'enrich' the water and feed algae, which then grow in masses that block out the sunlight and cause the death of water plants and creatures.
*Example*: Norfolk Broads.

**outcrop**: underlying rock that's exposed by weathering.
*Example*: The outcrops of granite on Dartmoor that form the TORS.

**oxbow lake**: curved lake, often the shape of a horseshoe, formed when a river cuts a shorter, straighter course through the neck of a MEANDER and isolates the former river bend.
*Example*: River Severn between A489 and B4569 at Caersws, Powys, mid-Wales. Oxbow lake in the process of being formed: Pennard Pill where it flows below Pennard Castle towards Three Cliffs Bay, Gower Peninsula, South Wales.

**peninsula**: land sticking out from a mainland mass into the sea.
*Example*: Llŷn Peninsula, north-west Wales.

**plain**: flat or gently rolling land, often grassland.
*Example*: Salisbury Plain, Wiltshire.

**plateau**: a high plain among mountains or hills.
*Example*: Cairngorm Plateau, among the mountains above the Day Lodge centre and car parks.

**point**: a headland.
*Example*: Morte Point, North Devon.

**port**: coastal facility to handle cargo and passenger ships.
*Example*: Port of Liverpool, Portpatrick (south-west Scotland), Port of London on the lower Thames.

**promontory**: a mass of land standing out over lower country-side, or sticking out into the sea.
*Example*: Buachaille Etive Mòr overlooking Glencoe, western Scotland; Knockdhu Promontory Fort, Co. Antrim; Filey Brigg, North Yorkshire coast; Great Orme, north Wales.

**raised beach**: beach left stranded high and dry by falling sea levels.
*Example*: west coast of Isle of Jura; in cliffs at Sewerby, East Yorkshire; between Rhossili Bay and Rhossili Down, Gower Peninsula, South Wales.

**rapids**: shallow water flowing quickly over and around rocks in a descending riverbed.
*Example*: River Wye at Symonds Yat; upper stretches of River Barle, Exmoor; Aysgarth Falls, North Yorkshire.

**reef**: rocks lying just below the sea surface.
*Example*: Eddystone Reef (supporting Eddystone Light-house), 10 miles (16 km) south-south-west of Plymouth Sound, Devon.

**reservoir**: manmade lake for storing water.
*Example*: Ladybower Reservoir, Derbyshire.

**ridge**: the crest of a hill or mountain that extends for some distance.
*Example*: Wenlock Edge, Shropshire; Hatterrall Ridge, Black Mountains, England/Wales border.

**river basin**: *see* CATCHMENT.

**rivulet**: small stream.
*Example*: Bourne Rivulet, near Andover, Hampshire (in its higher reaches).

**saltmarsh**: *see* MARSH.

**sandbank**: large area of sand, often tidal, off the coast.
*Example*: Off Lancashire coast around Southport; Goodwin Sands off Deal, Kent.

**scree**: loose stone forming an unstable sheet on a slope.
*Example*: Wastwater Screes, Lake District.

**sea stack**: freestanding pinnacle or block of rock, separated from mainland by the action of the sea.
*Example*: Old Man of Hoy, Isle of Hoy, Orkney Isles; Marsden Rock, off Tyne & Wear coast between Sunderland and South Shields.

**sea wall**: bank built along low-lying coast to prevent sea flooding.
*Example*: Canvey Island, Essex; North Norfolk and Lincolnshire coasts.

**silt**: tiny grains of soil or rock, carried and deposited by a river, that can choke off a harbour or port from the sea.
*Example*: North Norfolk coast; River Lune, Lancashire (when silt stopped ships travelling upriver to Lancaster, Glasson

Dock port was built near the river mouth and connected to the city by a canal).

**slag heap, spoil heap**: mound of waste material produced by mining or quarrying.
*Example*: Bersham Bank colliery tip near Wrexham, Clwyd, Wales; lead-mining heaps all over Yorkshire and Durham Dales.

**sound**: (a) a large sea inlet, or (b) a narrow channel of sea between two pieces of land.
*Example*: (a) Plymouth Sound; (b) Bluemull Sound between islands of Yell and Unst, Shetland.

**source** or **spring**: starting point of a river.
*Example*: The rivers Severn, Wye and Rheidol all have their sources on Plynlimon Mountain, west Wales.

**strait**: a narrow channel that connects two larger stretches of water.
*Example*: The Strait of Dover connects the North Sea with the English Channel.

**strath**: Scottish term for a wide valley, generally lower down a river than a GLEN.
*Example*: Strathspey, north-east Scotland.

**terminal moraine**: *see* MORAINE.

**tor**: rock outcrop at or near the top of a hill.
*Example*: Rough Tor and Shining Tor, near Camelford, Bodmin Moor, Cornwall.

**tributary**: smaller stream or river flowing into a bigger one.
*Example*: The River Pang is a tributary of the River Thames, flowing into it at Pangbourne, Berkshire.

**uplands**: an area of higher land, generally hilly rather than mountainous.
*Example*: The Southern Uplands region of the Scottish Borders.

**vale**: a wide river valley with a broad, flat FLOODPLAIN.
*Example*: Severn Vale between Bristol and Gloucester.

**volcanic plug**: formed when magma or molten rock solidifies in the vent or 'throat' of a volcano, and the rest of the volcano then erodes away, leaving the plug behind.
*Example*: Bass Rock, Firth of Forth near Edinburgh; Slemish Mountain, Co. Antrim, Northern Ireland.

**water meadow**: not a generic term for any grazing meadow by a river, as is often supposed, but a specially managed meadow where flooding is permitted because of the nutrient-rich silt it spreads, with channels and sluices to control the amount of water.
*Example*: Harnham Water Meadows, Salisbury, Wiltshire – scene of John Constable's painting, *Salisbury Cathedral from the Meadows* (1831).

**wolds**: low-rolling hills of CHALK or LIMESTONE, with little or no surface water.
*Example*: Cotswold Hills, Gloucestershire; Lincolnshire Wolds; Yorkshire Wolds.

## GEOLOGICAL TERMS

If Geography was a bore and a bafflement for me at school, Geology was infinitely worse: that parade of mind-crushingly meaningless dead stuff from way down who-cares-where?

Quartzite or chalk; dolerite or mudstone? Who gives a flying fault-zone metamorphism?

I do, that's who – now I've learned to look about me. Now I know (a) that the beautiful little spring sandwort flower likes lead in its lunch, (b) that flints are actually sheets of glass made from sea sponges, (c) that the ancient murder victims dug out of our bogs didn't rot away because the peat around them couldn't rot, (d) that the amazing U-shaped valley you see from High Cup on the Pennine Way is there because even a mile-high glacier couldn't crush the cliffs of dolerite, and (e) a whole lot of other Geology-related things that fascinate me. I realise that what's happened to the land over the past 3,000 million years has been, literally, the greatest story out there. It's still ongoing, too, viz. the eruption of the Icelandic volcano Eyjafjallajökull in April 2010, which stopped air traffic over northern Europe and gave us all a reminder that Nature is a whole hell of a lot more powerful than Man.

Here are the basic rock types I've mentioned in the book, and examples of where to see them at their best. Happy hunting!

**basalt**: a dark IGNEOUS rock, sometimes forming tall hexagonal columns like stacks of pencils. It weathers into tremendous ledges.
*Example*: The western Highlands and islands of Scotland, especially Mull, Staffa and the huge cliffs of Skye's Trotternish peninsula.

**carboniferous limestone**: *see* LIMESTONE.

**chalk**: a pale or white SEDIMENTARY rock, formed by the shells and skeletons of tiny plankton living and dying in the warm waters of the Great Chalk Sea, a shallow tropical ocean that covered most of Britain and north-west Europe around 60–80 million years ago.
*Example*: White Cliffs of Dover, Kent; Seven Sisters cliffs, East Sussex.

**clay**: composed of tiny grains of minerals produced by the slow weathering of a wide range of rocks, then transported and eventually laid down by water in the form of lakes or big, slow-flowing rivers. It's poor stuff for growing things in, but can be improved by adding organic matter such as compost.
*Example*: cliffs at Isle of Sheppey, Kent; The Naze, Essex.

**dolerite**: a hard IGNEOUS rock, a type of BASALT, dark in colour.
*Example*: The Whin Sill is a tongue of dolerite formed about 300 million years ago, which outcrops in spectacular cliffs at High Force and Cauldron Snout waterfalls in Upper Teesdale on the Durham/Cumbria border, along the line of Hadrian's Wall either side of Housesteads Roman Fort, and at Bamburgh Castle and the Farne Islands on the Northumberland coast.

**flint**: flint is formed of layers or nodules of a crystalline material called silica, produced by the decomposing skeletons of sea creatures such as sponges.
*Example*: on any beach below chalk cliffs, or in ploughed fields with chalky soil.

**freestone**: any stone used in masonry (but very often hard, marble-like LIMESTONE) that is fine-grained and soft enough

to cut with a chisel, yet capable of standing centuries of exposure and weathering.
*Example*: cliffs near Swanage on the Isle of Purbeck, Dorset; Isle of Portland, Dorset.

**gabbro**: a form of dark, coarse-grained BASALT that weathers into jagged 'teeth'.
*Example*: Black Cuillin mountains, Isle of Skye.

**gneiss**: this ancient type of GRANITE, often coarse-grained with big crystals embedded in it, is metamorphic.
*Example*: The isles of the Outer Hebrides, western Scotland, whose 'Lewissian gneiss' (found all over, not just on the Isle of Lewis) could be as much as 3,000 million years old. Think of that!

**granite**: a hard, IGNEOUS rock, pink, grey or black, or a speckly mixture, characteristically outcropping in tors like those on Dartmoor.
*Example*: West Cornwall; Bodmin Moor and Dartmoor; Eskdale, Lake District.

**greensand**: greenish-coloured SANDSTONE, often forming bands among CHALK and CLAY, deposited by ancient seas.
*Example*: All along the North and South Downs; Surrey Hills; Leith Hill; Devil's Punchbowl, Hindhead, Surrey.

**gritstone**: sometimes known as millstone grit (because it was perfect for making millstones) – a very coarse SANDSTONE which can contain quite large, quartz-like pebbles.
*Example*: Dark Peak of Derbyshire.

**igneous**: rock formed by molten lava cooling and becoming solid.
*Example*: basalt ledges of Isle of Mull, Inner Hebrides.

**ironstone**: a SEDIMENTARY, sandy rock, always weighing heavy in the hand because of its high iron content. It ages to a rusty, golden colour.
*Example*: Around Flockton, Emley and Cawthorne, West Yorkshire; south of Nettleton, Lincolnshire.

**limestone**: there are many types of limestone, a SEDIMENTARY rock that's often packed with fossils. All forms are rich in calcite (lime) and good for growing grass, flowers and trees. **Carboniferous limestone** was laid down under water in layers. When sea levels have fallen and left carboniferous limestone as dry land, rainwater widens the cracks between the layers and rivers eat away the rock into great caverns. **Oolitic limestone** is made of round grains, and forms a band of beautiful building stone, varying from dark gold to silver, that snakes right across England, from the Dorset coast north-east to the River Humber in Lincolnshire.
*Example*: Carboniferous limestone – Mendip Hills, Somerset. Oolitic limestone – Cotswold Hills, Gloucestershire.

**metamorphic**: rock that has been subjected to so much heat and pressure from volcanic processes that its chemical formula and/or physical properties have been changed.
*Example*: Quartzite outcrops of the Stiperstones, Shropshire.

**mudstone**: solid rock composed of clay or mud.
*Example*: Kimmeridge Bay, Dorset.

**oolitic limestone**: *see* LIMESTONE.

**quartzite**: METAMORPHIC rock, very hard and usually white or pink-coloured, which is basically sand squeezed and baked by enormous forces.

*Example*: Stiperstones, Shropshire; Holyhead Mountain on the Isle of Anglesey, north-west Wales; peaks of the Torridon Mountains, north-west Scotland.

**sandstone**: a SEDIMENTARY rock composed of grains of sand, mostly quartz and glassy crystals of the mineral feldspar.
*Example*: Red sandstone cliffs of Cumbria and west Dorset/east Devon.

**schist**: METAMORPHIC rock, often with its mineral grains elongated to make it quite easy to break, split along its many layers, and flake off. Schist generally starts as CLAY and/or mud, transformed by heat and squeezing into SHALE or SLATE, then continuing the process to form schist.
*Example*: Ben Lui, north-west of Loch Lomond; Ben Lawers, north of Loch Tay; Monadhliath Mountains, west of the Cairngorms; Ben More, south of the A85 in Glen Dochart.

**sedimentary**: rock formed by minute particles falling to the bottom of lakes, rivers or seas and piling up in layers.
*Example*: chalk cliffs of Beachy Head and the Seven Sisters, East Sussex.

**shale**: under pressure the microscopic CLAY particles that make up MUDSTONE can have their minerals consolidated into flat plates, easily split into the characteristic laminate sheets of shale.
*Example*: Kimmeridge Bay, Dorset, where the shale contains a kind of bitumen called kerogen, which can be heated to produce shale oil for use like petrol.

**slate**: grey or purple METAMORPHIC rock, generally produced when CLAY or volcanic ash are first heated and pressured

into SHALE, then into slate. It splits cleanly along its laminated planes, making it ideal for roofing materials.
*Example*: Honister Pass, Lake District; Dinorwig Quarries, Snowdonia.

# 11

## PUB QUIZ

'What two ideas are more inseparable than
Beer and Britannia?'
– Sydney Smith, clergyman and writer

The nice thing about this Pub Quiz is that you don't have to display your ignorance in front of a classroom full of sneering clever-clogses – not like those bad old days back in the Geography lesson. No – now you're (comparatively speaking) a grown-up, you can take it down to the pub with a pencil and noodle about with it on your own, or with a friend or two, over a couple of pints.

Just go back through the book before you start the quiz and check up on things you're not sure of.
The answers are at the end, and you could very easily cheat by looking at them as you go along, or by Googling the solutions. I probably would, in your shoes. But don't! Resist temptation and

YOU WILL BE A BETTER AND MORE GOOD-LOOKING PERSON BECAUSE OF IT.

Use a pencil, and then you can rub out your answers before you kindly wrap the book up and give it to someone else. OR ... photocopy the Pub Quiz and Answers on to some old A4 sheets and defile them as much as you like, having left my precious oeuvre in a warm, dry and safe place.

# 1. GREEDY GEOGRAPHER'S TOUR OF BRITAIN

Britain's famous for its local food (and drink). Grab your taste buds, and let's ride!

## Products and places

On the left are descriptions of forty delicious local UK products; on the right, forty numbered places. One point for correctly marrying up the product and the place associated with it.

OK, that's an easy free point – now try the rest ... (Answers at end of quiz.)

## Products

Angus beef

beef baked in puff pastry

bitter beer – hop flavour enhanced
    by local water

black pudding

blue-tinged cockles

bottled relish containing
    (among other things) anchovies,
    tamarinds, vinegar and sugar

brown ale

brown meat soup

cheese – crumbly, moist, pale, a
    good balance of honey and acid

cheese – crumbly, white, with a
    slightly sour tang

## Places

1. Aberdeen
2. Arbroath
3. Aylesbury,
    Buckinghamshire
4. Bakewell, Derbyshire
5. Banbury, Oxfordshire
6. Barnsley, S. Yorkshire
7. Bath, Somerset
8. Bedfordshire
9. Bournville,
    Birmingham
10. Brighton, Sussex
11. Burton, Staffordshire
12. Bury, Lancashire

cheese – hard, sharp, nutty, yellow

cheese – strong, savoury, made with
    full-cream milk, with a rich
    orange colour

cheese – strong-smelling, blue-veined, sharp

chocolate made by Quakers

double-sided lamb chop

duckling

edible crabs

flaky pastry cakes full of currants

flat round liquorice cakes

gin

kippers

long, curly sausage

meaty suet pudding known as a 'clanger'

menthol and eucalyptus lozenges – fishermen like them

mint cake

mustard

natural spring water

'Oliver' biscuits to eat with cheese

oval pastry cake with spices and currants

peppermint seaside rock

pork pies

portable snack, devised so the Earl could eat without leaving the gambling table

pudding (sometimes called 'tart') containing jam and almonds

13. Caerphilly, Wales
14. Cheddar, Somerset
15. Cromer, Norfolk
16. Cumberland (now part of Cumbria)
17. Dover, Kent
18. Dundee
19. Eccles, Lancashire
20. Edinburgh
21. Everton, Merseyside
22. Fleetwood, Lancashire
23. Gloucester, Gloucestershire
24. Jersey, Channel Islands
25. Kendal, Cumbria
26. Maldon, Essex
27. Malvern, Worcestershire
28. Melton Mowbray, Leicestershire
29. Newcastle-upon-Tyne, Tyne & Wear
30. Plymouth, Devon
31. Pontefract, W. Yorkshire
32. Sandwich, Kent
33. Stiffkey, Norfolk
34. Stilton, Cambridgeshire

rich fruit cake

'Royal' potatoes

sea salt

smoked haddock, known as 'smokies'

soft, crumbly, pastel-coloured sweet
    stick of rock

sole

stripy peppermint humbugs

35. Urchfont, Wiltshire
36. Wellington, Somerset
37. Wensleydale,
    N. Yorkshire
38. Whitby, N. Yorkshire
39. Windsor, Berkshire
40. Worcester,
    Worcestershire

# 2. BASICS

## Compass

**If the compass is a clock, and North is at noon, which compass directions are at:**

41. 3 o'clock?
42. 22½ minutes to the hour?
43. 7½ minutes past the hour?
44. 9 o'clock?

## Random

45. Which is generally drier, warmer and flatter – east or west side of Britain?
46. Which proportion of the UK population lives in England – 2/5ths, 5/6ths, 8/9ths?
47. Which Industrial Revolution product needed damp for its production?

[197

48. What is England's highest mountain?
49. And what is its height in feet or metres – 3,290 ft/1,003 m; 2,903 ft/885 m; 3,209 ft/978 m; 3,092 ft/942 m?
50. What is Scotland's highest mountain?
51. And its height – 4,094 ft/1,248 m; 4,249 ft/1,295 m; 4,940 ft/1,506 m; 4,409 ft/1,344 m?
52. And what is Wales's highest mountain?
53. And its height – 3,560 ft/1,085 m; 3,650 ft/1,113 m; 3,506 ft/1,066 m; 3,065 ft/934 m?
54. Name the largest lake in the British Isles.

## Basic Regions

Match the numbers to the circles on the map opposite.

55. Flow Country
56. Lake District
57. Powys
58. Lower Lough Erne
59. East Anglia
60. Scottish Lowlands
61. Yorkshire
62. Snowdonia
63. Lough Neagh
64. Great Glen
65. Welsh Valleys
66. London
67. Antrim Coast
68. Shakespeare Country
69. Cardiff
70. Scottish Highlands

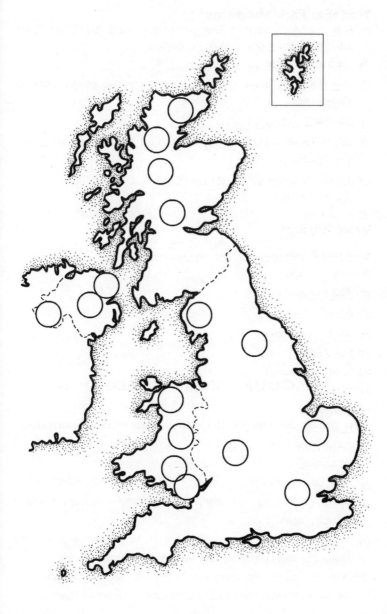

## National Park anagrams

71. At Mordor

72. Re bacon bonces

73. Man under brothl

74. Dislike tract

75. So drab

76. No grim scar

77. Distract pike

78. He is cats poem broker

79. Shrek rose daily

80. Rome ox

81. No sow thuds

82. Och no Moll D

83. And I swoon

84. Hot Nooky Mrs Orr

85. Worn feets

# 3. COUNTIES AND REGIONS

## Can you fit the county or Council Area to the description?

86. Landlocked county dotted with antiquities and centred on a vast emptiness.

87. This county mostly consists of lovely forested hills inland, with a motorway running west along the coast until it reaches industrial territory again.

88. It stretches from coast to coast across the slimmest bit of its country, and is the most populous, urbanised and industrialised part.

89. A flattish sweep of country. The North Sea eats away at its eastern section – crumbly black cliffs and sensational, deserted beaches.

90. All about limestone and lakes; one-third of it is under water. From the lakes you travel south through limestone country, which rain and rivers have eaten into spectacular caverns.

91. A jagged-edged scatter of peninsulas and islands poking out and down towards Northern Ireland.

92. Seaside holiday country, where the bucket and spade do more for the economy than the blast furnace and the coal shovel.

93. Home of the ancient and world-famous Quorn Hunt – a classic fox-hunting, horse-galloping county with its big, flattish fields.

94. Wide and wild moors with a huge wheel of empty upland – wonderful hiking country – at their heart. Out to the west, a flat coast of vast sandy beaches and wide muddy estuaries.

95. 'Bandit Country! Don't go there!' A tightly drawn and intimate dairy-farming landscape of small steep hills.

# 4. CITIES

## Nicknames

Which cities boast football clubs with these nicknames? Extra point if you can give the club's official name!

- The Hatchetmen
96. City:
97. Club:
- The Caley Jags
98. City:
99. Club:

- The Cabbage
100. City:
101. Club:
   - The Villans
102. City:
103. Club:
   - The Glorious Hoops
104. City:
105. Club:
   - The Jam Tarts
106. City:
107. Club:
   - The Toffees
108. City:
109. Club:
   - The Cocks 'n' Hens
110. City:
111. Club:
   - The Harry Wraggs
112. City:
113. Club:
   - The Valiants
114. City:
115. Club:

## Bands

Which cities are these bands associated with?

116. Wet Wet Wet
117. The Teardrop Explodes
118. Black Sabbath

119. Stiff Little Fingers
120. Super Furry Animals

## Famous folk

### Which cities are they associated with?

121. Dylan Thomas, poet
122. Stanley Matthews, footballer
123. Marie Stopes, birth control pioneer
124. Banksy, guerrilla artist
125. John Hume, politician and Nobel Peace Prize winner

## Go there for

### Where would you go for?

126. a stottie cake from Greggs the bakers
127. great pubs of Rose Street
128. great Indian food in the 'Balti Triangle'
129. giant cranes Samson and Goliath
130. art nouveau architecture of Charles Rennie Mackintosh

## Mad facts

### Which city?

131. has a Pizza Express that's haunted by a ghost?
132. produces 1.5 million tonnes of rubbish every year?
133. boasts the longest hospital corridor in Europe?
134. is haunted by Gabble Ratchets and Barguests?
135. has a football club so hard up that they cancelled their weekly order of clean jockstraps?
136. doesn't allow immodest and/or incontinent women within four miles?

# 5. WATERY BITS

## Stages of a river

137–150. Add the fourteen missing labels to this illustration of a river's stages.

## Rivers of Britain

Match the rivers to the lengths.

| | | |
|---|---|---|
| 151. | Tees | 5 miles (8 km) |
| 152. | Severn | 143 miles (230 km) |
| 153. | Bann | 220 miles (354 km) |
| 154. | Great Ouse | 70 miles (113 km) |
| 155. | Kervaig | 80 miles (129 km) |

. . . and to the sources . . .

| | | |
|---|---|---|
| 156. | near Ashe, Overton, Hampshire | Humber |
| 157. | Wrynose Pass, Furness Fells, Cumbria | Duddon |
| 158. | Watermeetings, Lowther Hills, South Lanarkshire | Tywi |
| 159. | confluence of River Trent and Yorkshire Ouse | Test |
| 160. | north of Llyn Brianne, near Llanwyrtyd Wells | Clyde |

## Courses

Which river flows via:

161. Cross Fell, High Force Waterfall, Barnard Castle, Middlesbrough?

162. Spelga Dam, Portadown, Lough Beg, Portglenone, Coleraine?

163. Cricklade, Abingdon, Wallingford, Maidenhead?

164. Garvamore, Kingussie, Charlestown of Aberlour, Fochabers?

165. Llanidloes, Welshpool, Bewdley, Tewkesbury, Sharpness?

# 6. BRITANNIA'S BULWARKS

*(Go on – have a go at some without any clues!)*

## Sea stretches

Which sea stretches would you look across if you were looking:

166. from the Lancashire coast west towards Ireland?

167. north from Duncansby Head towards the Orkney Isles?

168. from Exmoor north towards the Gower Peninsula?

169. from the Mull of Galloway west to Northern Ireland?

170. south from Dover towards France?

## Islands

Which islands are we talking about?

171. L-shaped, a pale 'cake' of chalky limestone topped with dark basalt 'icing', with a quarter of a million screaming seabirds.

172. An island with a split personality, half empty grazing marshes, half tight-packed housing, gas and fuel storage, and great R&B music.

173. Thirty times more red deer than humans, wild west coast of raised pebble beaches, quartzite peaks, George Orwell's home while he wrote *1984*.

174. Rocky, sheer, bare, bleak, never inhabited, utterly isolated in the Atlantic Ocean.

175. Off the tip of a peninsula, shaped like a sleeping otter, final resting place of countless medieval pilgrims.

# 7. LANDMARKS

176. A two-mile-long, green-backed promontory of humps, sticking out from the western tip of the Gower Peninsula like a monster putting out to sea.

177. Its three skeleton cantilevers are painted an eye-catching red. This proud piece of Victorian engineering doesn't want to blend into the low-lying, green and brown landscape — it's all about Man's dominion over Nature.

178. Britain's finest hill fort, visible for miles, created over 4,000 years from the Stone Age to the Iron Age. Multiple ramparts encircle the hilltop.

179. The volcanic plug of a mountain where St Patrick, patron saint of Ireland, herded pigs as a slave in his youth is a touchstone in Irish spirituality and a landmark for miles around in the flattish mid-Antrim landscape.

180. Is it a bird, is it a plane, is it a man with his arms strapped to an ironing board? No, it's a fabulous sculpture with the height of a tall factory chimney and the wingspan of a Boeing 767F, keeping watch from its knoll.

# 8. TERMS

## Match these terms and descriptions:

| | | |
|---|---|---|
| 181. boulder of an alien rock type, carried and deposited by a glacier | drumlin |
| 182. wide valley (Scottish) | tor |
| 183. loose stone forming an unstable sheet on a slope | beck |
| 184. Welsh valley | nant |
| 185. granite outcrop on Dartmoor | strath |

| | |
|---|---|
| 186. hill of gravel left by retreating glacier | lochan |
| 187. Northern English term for a stream | erratic |
| 188. small Scottish lakelet | cwm |
| 189. curve cut by a river through its floodplain | scree |
| 190. Welsh stream | meander |

# 9. GRAB-BAG

## Ten to finish:

191. What's the approximate length of Britain's coastline – 10,000 miles (15,000 km), 20,000 miles (30,000 km), or 30,000 miles (48,000 km)?

192. What was Swansea's nickname in the nineteenth century?

193. What kind of rock underlies the moors of the Dark Peak of Derbyshire?

194. Which city offers visitors a winter ale festival, a summer folk festival and an autumn festival of ideas?

195. What was the nickname of James Williamson, First Baron Ashton, who built the Ashton Memorial in Williamson Park, Lancaster, in memory of his wife?

196. Which jolly Scottish outlaw roamed the hills of the Trossachs?

197. How many Ice Ages (that we know about) has the world experienced?

198. What was the Scottish Borders' nickname before the Act of Union with England in 1707 brought some law and order?

199. How big is Lough Neagh on the CRCS (County of Rutland Comparison Scale)?

200. What has artist Serena de la Hey created on the Somerset Levels next to the M5?

# ANSWERS

Come on, now – you wouldn't sneak a crafty peep in your birthday parcel before the day itself, would you? Well then – get your nose out, and don't come back till you've tried 'em properly, eh?

# 1. GREEDY GEOGRAPHER'S TOUR OF BRITAIN

## Product

Angus beef (1)

beef baked in puff pastry (36)

bitter beer – hop flavour enhanced by local water (11)

black pudding (12)

blue-tinged cockles (33)

bottled relish containing (among other things) anchovies, tamarinds, vinegar and sugar (40)

brown ale (29)

brown meat soup (39)

cheese – crumbly, moist, pale, a good balance of honey and acid (37)

cheese – crumbly, white, with a slightly sour tang (13)

cheese – hard, sharp, nutty, yellow (14)

cheese – strong, savoury, made with full-cream milk, with a rich orange colour (23)

cheese – strong-smelling, blue-veined, sharp (34)

chocolate made by Quakers (9)

double-sided lamb chop (6)

duckling (3)

edible crabs (15)

flaky pastry cakes full of currants (19)

flat round liquorice cakes (31)

gin (30)

kippers (38)

long, curly sausage (16)

meaty suet pudding known as a 'clanger' (8)

menthol and eucalyptus lozenges – fishermen like them (22)

mint cake (25)

mustard (35)

natural spring water (27)

'Oliver' biscuits to eat with cheese (7)

oval pastry cake with spices and currants (5)

peppermint seaside rock (10)

pork pies (28)

portable snack, devised so the Earl could eat without leaving the gambling table (32)

pudding (sometimes called 'tart') containing jam and almonds (4)

rich fruit cake (18)
'Royal' potatoes (24)
sea salt (26)
smoked haddock, known as 'smokies'
   (2)

soft, crumbly, pastel-coloured
   sweet stick of rock (20)
sole (17)
stripy peppermint humbugs
   (21)

# 2. BASICS

## Compass

41. East
42. South-west

43. North-east
44. West

## Random

45. East
46. 5/6ths
47. Cotton
48. Scafell Pike
49. 3,209 ft/978 m

50. Ben Nevis
51. 4,409 ft/1,344 m
52. Snowdon
53. 3,560 ft/1,085 m
54. Lough Neagh

## Basic Regions

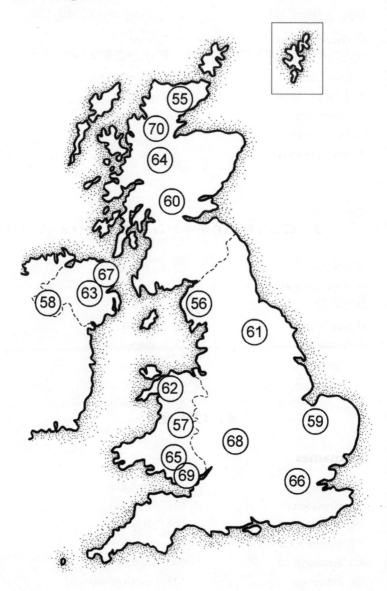

## National Park anagrams

71. Dartmoor
72. Brecon Beacons
73. Northumberland
74. Lake District
75. Broads
76. Cairngorms
77. Peak District
78. Pembrokeshire Coast

79. Yorkshire Dales
80. Exmoor
81. South Downs
82. Loch Lomond
83. Snowdonia
84. North York Moors
85. New Forest

# 3. COUNTIES AND REGIONS

86. Wiltshire
87. West Glamorgan
88. Central
89. East Riding
90. Fermanagh

91. Argyll & Bute
92. Dyfed
93. Leicestershire
94. Lancashire
95. Armagh

# 4. CITIES

## Nicknames

96. Belfast
97. Crusaders
98. Inverness
99. Inverness Caledonian Thistle
100. Edinburgh
101. Hibernian

102. Birmingham
103. Aston Villa
104. Glasgow
105. Queen's Park
106. Edinburgh
107. Heart of Midlothian

108. Liverpool
109. Everton
110. Belfast
111. Glentoran

112. Glasgow
113. Partick Thistle
114. Stoke-on-Trent
115. Port Vale

## Bands

116. Glasgow
117. Liverpool
118. Birmingham

119. Belfast
120. Cardiff

## Famous folk

121. Swansea
122. Stoke-on-Trent
123. Edinburgh

124. Bristol
125. Derry

## Go there for

126. Durham
127. Edinburgh
128. Birmingham

129. Belfast
130. Glasgow

## Mad facts

131. Inverness
132. Manchester
133. Cardiff

134. Leeds
135. Portsmouth
136. Cambridge

# 5. WATERY BITS

## Stages of a river

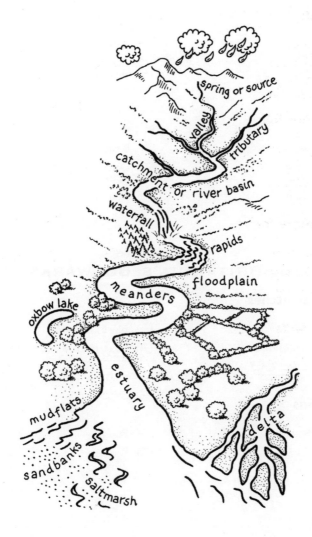

## Rivers of Britain

### Lengths

151. 70 miles (113 km)
152. 220 miles (354 km)
153. 80 miles (129 km)

154. 143 miles (230 km)
155. 5 miles (8 km)

### Sources

156. Test
157. Duddon
158. Clyde

159. Humber
160. Tywi

### Courses

161. Tees
162. Bann
163. Thames

164. Spey
165. Severn

# 6. BRITANNIA'S BULWARKS

## Sea stretches

166. Irish Sea
167. Pentland Firth
168. Bristol Channel

169. North Channel
170. Strait of Dover

## Islands

171. Rathlin Island
172. Canvey Island
173. Jura

174. Rockall
175. Bardsey

# 7. LANDMARKS

176. Worm's Head
177. Forth Railway Bridge
178. Maiden Castle

179. Slemish
180. *Angel of the North*

# 8. TERMS

181. erratic
182. strath
183. scree
184. cwm
185. tor

186. drumlin
187. beck
188. lochan
189. meander
190. nant

# 9. GRAB-BAG

Ten to finish . . .

191. 20,000 miles (30,000 km)
192. Copperopolis
193. Gritstone
194. Cambridge
195. The Oilcloth King

196. Rob Roy
197. Five
198. The Debatable Lands
199. Same size
200. Willow Man

## And Now . . .

Add 'em up!

   0–50: Well . . . there's always poetry.

 51–100: Hmmm.

101–150: Pretty good. Another pint and you'll be there.

151–190: A triumph! You are a SPLENDID
         EXAMPLE to us all.

191–200: Bloody cheat.

# 12

## NOW WASH YOUR HANDS

That's it, folks!

It's been a great cathartic pleasure writing this book and exorcising that old demon, Geography. Actually Geography has become a guiding light and a fountainhead of pleasure over the course of the years – probably because I've shied away from the label, and encountered it instead as 'fabulous places for walks', or 'gorgeous landscape', or 'flowers that always grow here and never over there', or simply 'the story of our countryside'. It *is* a story, a brilliant and ongoing one. I'm glad I've finally started it. How things will turn out for our hero, the little archipelago with the big character, is anybody's guess – that's still way in the future, a few gazillion years down the line.

Thanks, Mr Matt, for doing your best with me back in the schoolroom. I didn't appreciate a bloody word of it then, but I do now. I'll wander and explore and puzzle out the Geography of these endlessly fascinating islands till the day I die, and I'll always have the hag and the pig, the parrot and the shredded wheat in the backpack. They might yet come in handy.

## Acknowledgements

First and foremost my very grateful thanks are due to Rupert Lancaster of Hodder & Stoughton for coming up with the idea of an 'Everyman's Geography', and to Claire Littlejohn for her wonderful illustrations.

I have great pleasure in acknowledging the help and support of these lovely people and organisations:

- my parents John and Elizabeth Somerville for giving me a childhood in the strange and exciting surroundings of the often-flooded Vale of the River Severn, and for allowing me the freedom to go out and explore it far from adult supervision – the best possible way to develop and learn to love a personal geography

- a succession of editors at *The Times, Daily Telegraph, Irish Independent* and other newspapers and magazines, with the generosity and insight to encourage my writing about the odd corners and back country bits of the British Isles

- the rangers and wardens of the National Trust, Natural England, Wildlife Trusts, RSPB and other countryside agencies throughout these islands who share their expertise so generously and enthusiastically

- Ordnance Survey for their incomparable maps, the best in the world, that bring our islands' geography alive and make it accessible to anyone with a pair of boots

- Alfred Wainwright, Seamus Heaney, Henry Williamson, Robert Macfarlane, Emily Brontë, Antony Gormley, Wilko Johnson, Sean Street, Fay Godwin, Jack Yeats and their fellow poets, painters, sculptors, photographers and writers who impel us to go out and find the landscapes that inspired them

- John Francis, Tony Aldous and Marcus Elsegood; Dave Richardson and Andy Lyddiatt; Olcan Masterson and Michael Gibbons; the members of Bristol's illustrious Club Foot; and others who have joined me in exploring the geography of these islands through walking, music and talk

I'm very grateful, as always, to my wife Jane, who supports and encourages me all the while, and loves to walk the landscapes of England, Scotland, Wales and Ireland with me, sharing her delight in everything and making me see all the things I'd otherwise miss.

# INDEX